Magnetic Memory

If you are a semiconductor engineer or a magnetics physicist developing magnetic memory, get the information you need with this, the first book on magnetic memory.

From magnetics to the engineering design of memory, this practical book explains key magnetic properties and how they are related to memory performance, characterization methods of magnetic films, and tunneling magnetoresistance effect devices. It also covers memory cell options, array architecture, circuit models, and read-write engineering issues.

You'll understand the soft-fail nature of magnetic memory, which is very different from that of semiconductor memory, as well as methods to deal with the issue. You'll also get invaluable problem-solving insights from real-world memory case studies.

This is an essential book, both for semiconductor engineers who need to understand magnetics, and for magnetics physicists who work with MRAM. It is also a valuable reference for graduate students working in electronic/magnetic device research.

Denny D. Tang is Vice President of MagIC Technologies, Inc., and has over 30 years' experience in the semiconductor industry. After receiving his Ph.D. in Electrical Engineering from the University of Michigan in 1975, he spent 15 years at IBM T. J. Watson Research Center, Yorktown Heights, NY, 11 years at IBM Almaden Research Center at San José, CA, and 6 years at Taiwan Semiconductor Manufacturing Company (TSMC). He is a Fellow of the IEEE, TSMC, and the Industrial Technology Research Institute (ITRI).

Yuan-Jen Lee is a Senior Engineer at MagIC Technologies, Inc., where he develops advanced magnetic memory technology. He received his Ph.D. from the National Taiwan University in 2003, after which he worked for ITRI, Hsinchu, Taiwan, developing toggle MRAM and spin-torque MRAM.

Magnetic Memory

Fundamentals and Technology

DENNY D. TANG AND YUAN-JEN LEE

MagIC Technologies, Inc.

CAMBRIDGE
UNIVERSITY PRESS

CAMBRIDGE
UNIVERSITY PRESS

Shaftesbury Road, Cambridge CB2 8EA, United Kingdom

One Liberty Plaza, 20th Floor, New York, NY 10006, USA

477 Williamstown Road, Port Melbourne, VIC 3207, Australia

314–321, 3rd Floor, Plot 3, Splendor Forum, Jasola District Centre, New Delhi – 110025, India

103 Penang Road, #05–06/07, Visioncrest Commercial, Singapore 238467

Cambridge University Press is part of Cambridge University Press & Assessment, a department of the University of Cambridge.

We share the University's mission to contribute to society through the pursuit of education, learning and research at the highest international levels of excellence.

www.cambridge.org
Information on this title: www.cambridge.org/9780521449649

First published 2010

A catalogue record for this publication is available from the British Library

Library of Congress Cataloging-in-Publication data
Tang, Denny D.
Magnetic memory : fundamentals and technology / Denny D. Tang, Yuan-Jen Lee.
    p. cm.
  ISBN 978-0-521-44964-9 (Hardback)
  1. Magnetic memory (Computers) I. Lee, Yuan-Jen. II. Title.
  TK7895.M3T36 2010
  621.39′73–dc22

                                      2009051398

ISBN   978-0-521-44964-9   Hardback

Contents

Preface

The advent of semiconductor technology has impacted the lives of many of us since the 1970s. Silicon CMOS (complementary metal-oxide-semiconductor) devices are practically ubiquitous, and by the year 2000, the value of the semiconductor industry exceeded that of the automobile industry. The magnetic industry, on the other hand, is much smaller than the semiconductor industry. Engineering schools of universities rarely cover any courses in this discipline. Nonetheless, a tiny magnetic recording device is in the hard disk of every computer. Like CMOS devices, magnetic recording technology is being scaled down from generation to generation. At the time of writing, the physical size of the magnetic bit remains smaller than a DRAM bit on silicon chips.

Researchers working in these two communities had little in common until the development of the modern magnetic random access memory, or MRAM. A MRAM chip is built by integrating magnetic tunneling junction (MTJ) devices onto the silicon CMOS circuits. The research activity of MTJs in academia and industry, both hard disk and semiconductor, has been very active since it first showed signs of technology implication in the mid 1990s. That effort led to the mass production of the MTJ recording head in hard disk in 2006. In the same year, the semiconductor industry announced the first successful introduction of an MTJ memory product. The viability of MTJ technology is proven. It is expected that research activities will develop further, which will increase cooperation between these two research communities. The purpose of this book is to facilitate the dialog and to bridge the gap. Each simple homework problem and answer is designed to help readers to link the magnetics to the memory performance. Thus, the book is suitable for those with discipline of semiconductor devices and wish to expand their knowledge base into the field of magnetic memory, and for those in magnetics who wish to "fine-tune" magnetics for MRAM chips.

The book is organized into seven chapters. Chapter 1 reviews the electric current, as most electrical engineering students learn, in their sophomore and junior years, that magnetism results from an electric current. This chapter introduces readers to the unit conversion ready for the discussion in Chapter 2, which deals with the origin of magnetism in materials and introduces the concepts of electron spin, magnetic moments and its dynamics. It covers the microscopic view of the magnetic moment of an electron and an atom, and investigates its relationship with the macroscopic properties of magnetic thin film materials. Once the film is patterned to make devices, it behaves very differently from a full film. Chapter 3 covers the properties of the patterned thin magnetic films. This leads to the discussion of magnetization switching properties of many modern magnetic RAM devices. Chapter 4 introduces the magnetoresistance effect in thin film stacks, covering AMR (anisotropy magneto-resistance), GMR (giant magneto-resistance) and TMR (tunnel magneto-resistance) effects. The magneto-resistance effect is the operational principle of all modern non-volatile magnetic memories. A thorough discussion of the magnetic tunnel junction is presented. A detailed description of the properties and the design of field-write modes magnetic memory device are given in Chapter 5 and that of spin-torque transfer mode in Chapter 6. The discussion also covers circuit aspects of the memory cell and memory array, and the circuit model of the magnetic tunnel junction device, so that one can gain a better perspective of the merits in the design of the magnetic tunnel junction for memory. Chapter 7 covers the present memory market and the position of the magnetic memory in this market. New applications of this technology will also be discussed.

This is a very active field. Papers and patent applications of the related subject appear continuously and in large quantities. This book aims to provide the reader with a sufficient understanding of the fundamental physics of magnetics, the properties of magnetic thin film materials, device properties, design, memory operation and many other aspects of engineering. It also aims to give those working with semiconductors a head start so that they may bring in more fruitful results to this relatively new field.

Acknowledgments

The authors would like to acknowledge the contribution made by their colleagues at MagIC Technologies, Inc., including Mao-min Chen, Terry Thong, Wiltold Kula, Cheng Horng, Ruth Tong, David Heim, Tai Min, Robert Beach, Guenole Jan and Karl Yang. Parts of their work are described in this book.

In addition, the authors would like to thank Stuart Parkin, William Gallagher, Jonathan Sun and Daniel Worledge of IBM; Professor Ching-Ray Chang of National Taiwan University; C.T. (Jack) Yao and Pantas Sutardja for numerous interactions; Anthony Oates of TSMC and Tak Ning of the IBM Watson Research Center, who helped us begin the writing process. The authors would also like to thank their families for lending their support during the manuscript preparation, especially Grace Tang, Pi-ju Liao and Minchene Tang.

Special thanks go also to our colleague in MagIC Technologies, Inc., Dr. Pokang Wang, for his critical reading of the manuscript and suggestions.

1   Basic electromagnetism

1.1   Introduction

Two thousand years ago, the Chinese invented the compass, a special metallic needle with one end always pointing to the North Pole. That was the first recorded human application of magnetism. Important understandings and developments were achieved in the mid 19th century and continue into the present day. Indeed, today, magnetic devices are ubiquitous. For example, to name just two: energy conversion devices provide electricity to our homes and magnetic recording devices store data in our computers. This chapter provides an introduction to basic magnetism. Starting from the simple attractive (or repelling) force between magnets, we define magnetic field, dipole moment, torque, energy and its equivalence to current. Then we will state the Maxwell equations, which describe electromagnetism, or the relationship between electricity and magnetism.

A great tutorial is provided by Kittel [1], which may be used to support students studying Chapters 1–4.

1.2   Magnetic forces, poles and fields

In the early days, magnetic phenomena were described as analogous to electrical phenomena: like an electric charge, a magnetic pole was considered to be the source of magnetic field and force. The magnetic field was defined through the concept of force exerted on one pole by another. In cgs units, the force is proportional to the strength of the magnetic poles, defined as

$$F = \frac{p_1 p_2}{r^2},\tag{1.1}$$

where r is the distance between two poles (in units of centimeters) and the unit of force F is the dyne. There is no unit for the pole, p. This equation defines the pole strength as one unit, when the force is 1 dyne and the distance between the poles is 1 cm. Analogous to Coulomb's Law of electric charge, one may consider a magnetic pole, say p_1, which generates a magnetic field H, and H exerts a force on the other pole, p_2. Thus,

$$F = \left(\frac{p_1}{r^2}\right) p_2 = H\, p_2,\tag{1.2}$$

where H is given by

$$H = \frac{p_1}{r^2}.$$ (1.3)

Thus, a magnetic field H of unit strength exerts a force of 1 dyne onto one unit of magnetic pole. The unit of the magnetic field in cgs units is the oersted (Oe). To get a feel for the strength of the magnetic field, at the end of a magnetic bar on a classroom white board the magnetic field can be as high as 5000 Oe, whereas the earth's magnetic field is smaller than 0.5 Oe.

1.3   Magnetic dipoles

Although a magnetic pole is the counterpart of an electric charge, there is a difference. Magnetic poles always come in pairs: a north pole and a south pole. A monopole has never been found. This pair of positive and negative poles occurs at the same time and forms a dipole. For example, a bar magnet always has a north pole at one end and a south pole at the other. Magnetic field lines emit from one pole, diverge into the surroundings and then converge and return into the other pole of the magnet. Figure 1.1 shows the field lines around a magnet.

If a bar magnet with a north pole and a south pole is dissected into two bars, will these two bars becomes two magnets? The answer is yes, since poles are always in pairs.

In 1820, H. C. Oersted discovered that a compass needle could be deflected when electric current passes through a wire positioned near to the compass. This was the first time electricity was linked to magnetic phenomena. Subsequent work by André-Marie Ampère established the basis of modern electromagnetism. He established the relationship between a magnetic dipole and a circulating current in a conductor loop around an axis. The direction of the dipole is along the axis of the loop, which is orthogonal to the loop plane. Figure 1.2 illustrates

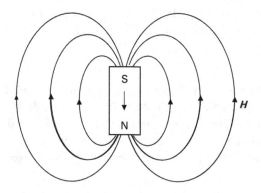

Figure 1.1. Magnetic field lines from a magnet.

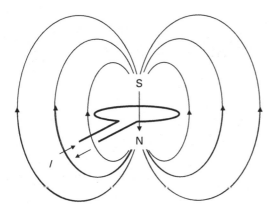

Figure 1.2. Magnetic field from a circular current loop.

the relationship between the dipole and the current loop. The polarity is dictated by the direction of the current, I. Reversing the direction of the electric current changes the polarity of the dipole. Thus, the magnetic dipole is another form of electric current, or moving electric charge.

Although both the electric field and the magnetic field are originated from electric charges, the difference is that the magnetic field must come from moving electric charges or electric current, rather than a stationary electric charge. The moving electric charge concept adequately explained the origin of magnetic poles at the time. The observation was later proven incorrect, however, when electron spin was taken into consideration. We will discuss this topic in a later part of this book.

1.4　Ampère's circuital law

Ampère further established that the relationship between the magnetic field H and the current I is given by

$$\oint H \cdot dl = 4\pi \cdot 10^{-4} I \tag{1.4a}$$

where, in cgs units, H is in oersted, dl is in centimeters and I is in milliamperes, and later one finds that it is more convenient to calculate H in thin films using

$$\oint H \cdot dl = 4\pi I \tag{1.4b}$$

(H is in oersted, dl is in micrometers and I is in milliamperes), where dl is the segment length of an arbitrary closed loop where the integration is performed and I is the current within the closed loop. This law is simple in concept and is

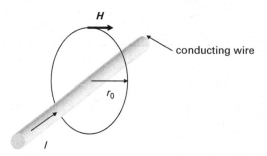

Figure 1.3. Magnetic field around a conducting wire carrying a current I. The magnetic field at a distance r_0 from the wire is $H = I/2\pi r_0$.

particularly useful in computing the field generated by the current in a long conductor and conducting thin film.

Here, we would like to discuss the units. Historically, there have been two complementary ways of developing the theory and definitions of magnetism. As a result, there are two sets of units for magnetic field, and thus for a magnetic pole. The definitions are similar, but not entirely identical. The major difference lies in how the magnetic field is defined inside the material. Centimeter-gram-second (cgs) units are used for studying physics, such as the origin of the magnetic pole and the magnetic properties in a material. The Systéme International d'Unités (SI units) are frequently used for obtaining magnetic field strength from circulating currents. Engineers working on electromagnetic waves, electric motors, etc. like to use SI units. This book will use both sets of units, depending on whichever makes more sense and in line with journal publications.

In SI units, Ampère's law is given as

$$\oint \boldsymbol{H} \cdot d\boldsymbol{l} = I. \tag{1.5}$$

(In SI units, $\boldsymbol{H}$ is given in amperes/meter, $d\boldsymbol{l}$ is in meters and I is in amperes.) From these two equations, one finds that a magnetic field of 1 (Oe) $= 1000/4\pi$ (A/m) ~ 80 (A/m).

Example 1.1: The magnetic field lines go around a current-carrying wire in closed circles, as illustrated in Fig. 1.3. At a distance r_0 from the conductor, the magnitude of the field $\boldsymbol{H}$ is constant. This makes the line integral of Ampère's law straightforward. It is simply given by

$$\oint \boldsymbol{H} \cdot d\boldsymbol{l} = 2\pi r_0 \boldsymbol{H} = I,$$

and so the field $\boldsymbol{H}$ is given by

$$H = \frac{I}{2\pi r_0}.$$

1.5   Biot–Savart Law

An equivalent statement to Ampère's circuital law (which is sometimes easier to use for certain systems) is given by the Biot–Savart Law. The Biot–Savart Law states that the fraction of a field, δH, is contributed by a current I flowing in an elemental length, δl, of a conductor:

$$\delta H = \frac{1}{4\pi r^2} I \, \delta l \times n, \tag{1.6}$$

(in SI units), where r is the radial distance from the current element and n is a unit vector along the radial direction from the current element to the point where the magnetic field is measured. Note that the direction of the vector δH is orthogonal to the plane formed by $I \, dl$ and n, as a result of the vector operation "$\times$" of two vectors $I \, dl$ and n, and the amplitude of $|I \, dl|\sin \theta$, where θ is the angle between vectors dl and n.

Example 1.2: Field from a current in a loop wire The magnetic field at the center of the loop plane as shown in Fig. 1.2 is calculated by the Biot–Savart Law as follows.

The radius of the loop is r_0, and H can be in the positive or negative z-direction, depending on the current direction, and only in the z-direction. The vector sum is simplified into a scalar sum. On the loop plane, $z = 0$. So, $|H| = H_0$, where H_0 is the integral of the field contributed by each segment dl of the loop, and

$$H_0 = 2\pi r_0 \left[\frac{1}{4\pi r_0^2} I \right] = \frac{1}{2r_0}, \tag{1.7}$$

(in SI units) and

$$H = H_0 n_z,$$

where n_z is the unit vector in the z-direction.

1.6   Magnetic moments

Next we need to introduce the concept of magnetic moment, which is an angular moment exerted on either a bar magnet or a current loop when it is in a magnetic field. The angular moment causes the dipole to rotate.

For a bar magnet positioned at an angle ϕ to a uniform magnetic field, H, as shown in Fig. 1.4, the forces on the pair of poles are given by $F_+ = +pH$ and $F_- = -pH$. The two forces are equal but have opposite direction. So, the moment acting on the magnet, which is just the force times the perpendicular distance from the center of the mass, is

$$pH \sin \phi (l/2) + pH \sin \phi (l/2) = pH \, l \sin \phi = mH \sin \phi, \tag{1.8}$$

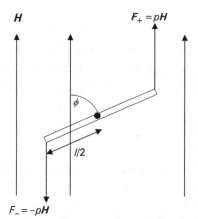

Figure 1.4. Dipole moment of a bar magnet in a uniform magnetic field.

where $m = pl$, the product of the pole strength and the length of the magnet, is the amplitude of the magnetic moment. The magnetic moment is a vector, pointing to a direction normal to the plane formed by the magnet and the magnetic field. One cgs unit of magnetic moment is the angular moment exerted on a magnet when it is perpendicular to a uniform field of 1 Oe. The cgs unit of magnetic moment is the emu (electromagnetic unit).

Since a magnetic dipole is equivalent to a current loop, it can be quantified by loop area A and a current I in the loop, and its magnetic moment is defined as

$$m = IAn, \tag{1.9}$$

where n is a vector normal to the plane of the current loop. In SI units, magnetic moment is measured in amperes times squared meters $(A\,m^2)$.

1.7　Magnetic dipole energy

A magnetic dipole can be defined in two ways. First, it is the magnetic moment, m, of a bar magnet at the limit of very short but finite length. Second, it is the magnetic moment, m, of a current loop at the limit of a very small but finite loop area. Either way, there is a finite magnetic moment.

The energy of a magnetic dipole is defined to be zero when the dipole is perpendicular to a magnetic field. So the work done in turning through an angle ϕ against the field is given by

$$\delta E = 2(pH \sin \phi)(l/2)\, d\phi = mH \sin \phi\, d\phi,$$

and the energy of a dipole at an angle ϕ to a magnetic field is given by

$$E = \int_{\pi/2}^{\phi} mH \sin \phi\, d\phi = -mH \cos \phi = -m \cdot H \tag{1.10}$$

This expression for the energy of a magnetic dipole in a magnetic field is in cgs units. In Eq. (1.10), E is in erg, m is in emu and H is in Oe. Equation (1.10) is also known as the formula for magnetostatic energy. In SI units the energy is $E = -\mu_0 m \cdot H$. When the dipole moment, m, is in the same direction as H, the magnetostatic energy takes its lowest value.

The torque exerted on a dipole moment is the gradient of the dipole energy with respect to the angle ϕ, or

$$\Gamma = dE/d\phi = mH \sin \phi. \tag{1.11}$$

The torque is exerted in the direction that lowers the dipole energy and the unit is expressed in erg/radian. When m and H are parallel, or $\phi = 0$, the energy is at a minimum, and the torque is zero. The torque is maximum when $\phi = \pi/2$. We will be using the concept of magnetic dipoles, and this expression for its energy in a magnetic field is used extensively throughout this book.

1.8 Magnetic flux

Here, we introduce another parameter: the flux Φ. Flux is defined as the integrated strength of a normal component of magnetic field lines crossing an area, or

$$\Phi = \int (H \cdot n) dA, \tag{1.12}$$

where n is the unit vector normal to the plane of the cross-sectional area, A. In cgs units, the flux is expressed in oersted times squared centimeters (Oe cm^2).

Magnetic flux is an important parameter in electric motor and generator design. The time-varying flux induces an electric current in any conductor which it intersects. Electromotive force ε is equal to the rate of change of the flux linked with the conductor:

$$\varepsilon = -\frac{d\Phi}{dt}. \tag{1.13}$$

This equation is Faraday's Law of electromagnetic induction. The electromotive force provides the potential difference that drives the electric current in a conductor. The minus sign indicates that the induced current sets up a time-varying magnetic field that acts against the change in the magnetic flux. This is known as Lenz's Law. The units in Eq. (1.13) as expressed in SI are: flux in webers (Wb), time in seconds and an electromotive force in volts.

1.9 Magnetic induction

When a magnetic field, H, is applied to a material, the response of the material to H is called magnetic induction, B. The relationship between B and H is a property of the material. In some materials (and in free space) B is a linear function of H. But in general B saturates at high H field and sometimes B is history-dependent and multiple-valued for each value of H. The equation relating B and H is given by

$$B = H + 4\pi M, \tag{1.14}$$

where M is the magnetization of the medium and B and H are given in cgs units gauss and oersted, respectively. The magnetization is defined to be the magnetic moment per unit volume:

$$M = m/V \tag{1.15}$$

in units of emu/cm^3. Note that M is a property of the material that depends on both the individual magnetic moments of the constituent ions, atoms and molecules, and on how these dipole moments interact with each other. Note that, although M is expressed in emu/cm^3, the unit of $4\pi M$ in Eq. (1.14) is not emu/cm^3, but gauss. See [2, 3] for good reference articles. In a vacuum, the magnetic induction B equals the magnetic field H since $M = 0$. Thus, 1 Oe field induces 1 gauss induction in a vacuum.

In SI units, the relation between B, H and M is given by

$$B = \mu_0(H + M), \tag{1.16}$$

where μ_0 is the permeability of free space. The units of M are obviously the same as those of H (A/m), and those of μ_0 are Wb/(A m), also known as henry/m. So the units of B are Wb/m^2, or tesla (T), and 1 gauss $= 10^{-4}$ tesla.

The magnetic induction, B, is the same thing as the density of flux, Φ, inside the medium. So, $B = \Phi/A$ in a material, by analogy with $H = \Phi/A$ in free space. In general, the flux density inside a material is different from that outside. In fact, magnetic materials can be classified based on the difference between their internal and external flux.

1.10   Classical Maxwell equations of electromagnetism

The magnetic dipole is the product of a circulating charged particle around an axis. The circulating charged particle around an axis forms a circulating electric current. Thus, the magnetic dipole is another form of electric current. Therefore, magnetic and electric phenomena have the same origin. Maxwell studied their relationship and elegantly described them in the four classical Maxwell equations of electromagnetics in SI units as follows [4]:

$$\nabla \cdot D = \rho, \tag{1.17}$$

$$\nabla \cdot B = 0, \tag{1.18}$$

$$\nabla \times H = J_c + dD/dt, \tag{1.19}$$

$$\nabla \times E = -dB/dt. \tag{1.20}$$

They are frequently brought together with the current charge relation:

$$\nabla \cdot J_c = -d\rho/dt. \tag{1.21}$$

The mathematical operator ∇ is called *del*; $\nabla\cdot$ means *divergence*. When operated on a vector D, it means the *divergence of vector D*. The result is a scalar, not a vector. In an (x, y, z)-coordinate system, it can be expressed as

$$\nabla \cdot D = \frac{\partial D_x}{\partial x} + \frac{\partial D_y}{\partial y} + \frac{\partial D_z}{\partial z}. \tag{1.22}$$

The notation $\nabla\times$ means *curl*. When operated on a vector H, it means the *curl of vector H*. The result is also a vector. In an (x, y, z)-coordinate system, it can be expressed as

$$\nabla \times H = \begin{vmatrix} n_x & n_y & n_z \\ \frac{\partial}{\partial x} & \frac{\partial}{\partial y} & \frac{\partial}{\partial z} \\ H_x & H_y & H_z \end{vmatrix} = \left(\frac{\partial H_z}{\partial y} - \frac{\partial H_y}{\partial z}\right) n_x - \left(\frac{\partial H_x}{\partial z} - \frac{\partial H_z}{\partial x}\right) n_y$$
$$+ \left(\frac{\partial H_y}{\partial x} - \frac{\partial H_x}{\partial y}\right) n_z, \tag{1.23}$$

where n_x, n_y, n_z are unit vectors in the x-, y- and z-directions, respectively.

The first equation, Eq. (1.17), shows that electrical flux D diverges out from an electric charge ρ. The second equation, Eq. (1.18), tells us that magnetic induction B does not diverge. Thus, B is continuous, forming a closed loop. There is no isolated magnetic charge, or pole, like there is in electric charge. Rather, the source of H is a current element J_c or a time-varying electrical flux D, as described in Eq. (1.19)! The current element J_c acts like a magnetic dipole. The curl of H indicates that the direction of the magnetic field H is orthogonal to the direction of J_c. For example, if J_c is in the z-direction, $\nabla \times H$ has no component in the z-direction, or H is in the x-y plane. This is another way to describe Ampère's Law.

Equation (1.20) demonstrates that the electric field E is induced by a time-varying magnetic induction B. This is the principle of electric generators and electric motors. Although Eq. (1.21) is not one of the Maxwell equations of electromagnetism, it indicates that current J_c can be viewed as the moving electric charge ρ. Together, these equations are the basis of electromagnetic wave propagation and energy conversation.

Since the Maxwell equations do not explicitly link material parameters and the behavior of magnetism, they are not commonly utilized in analyzing the behavior of magnetic materials and thin films.

1.11   Inductance

A time-varying current I in a conducting wire will generate a time-varying magnetic induction B around the wire. According to the Maxwell equation (1.20), the time-varying B induces an electric field in the wire. Thus, a voltage appears at the

Table 1.1 Cgs to SI parameter table

	cgs	SI
Force between poles	$F = \dfrac{p_1 p_2}{r^2}$ (dyne)	$F = \dfrac{1}{4\pi\mu_0} \dfrac{p_1 p_2}{r^2}$ (N)
Field of a pole	$H = \dfrac{p}{r^2}$ (Oe)	$H = \dfrac{1}{4\pi\mu_0} \dfrac{p}{r^2}$ (A/m)
Magnetic moment	$m = p(length)$ (emu)	$m = A I$ (A · m^2)
Magnetization	$M = m/(volume)$ (emu/cm^3)	M (tesla)
Magnetic induction	$B = H + 4\pi M$ (gauss)	$B = \mu_0 H + M$ (tesla)
Energy of a dipole	$E = -m \cdot H$ (erg)	$E = -\mu_0 m \cdot H$ (J)
Magnetic susceptibility	$\chi_m = \dfrac{M}{H}$ (emu/cm^3/Oe)	$\chi_m = \dfrac{M}{H}$ (dimensionless)
Permeability	$\mu = \dfrac{B}{H} = (1 + 4\pi\chi_m)$ (gauss/Oe)	$\mu = \dfrac{B}{H} = \mu_0(1 + \chi_m)$ (henry/m)

two ends of the wire. In electronic circuitry, the relationship between the induced voltage V and the time-varying current I in a conducting wire is given by

$$V = L\frac{dI}{dt},$$ (1.24)

where L is called the *inductance*. The unit of L is the (volt · second)/ampere or the *henry* (H); L is also called the *self-inductance*, since the time-varying current in the wire itself is the source of the voltage across the two ends of the wire through magnetic induction.

Equation (1.20) does not restrict the source of time-varying B; this equation also allows a time-varying I_1 in the first wire to produce a time-varying B, which extends to a neighboring second wire and induces a voltage V_2 in the neighboring second wire. The relation between the current and induced voltage is given by

$$V_2 = L_{21}\frac{dI_1}{dt},$$ (1.25)

where L_{21} is called *mutual inductance*.

Inductance is an electronic circuit element, like resistance and capacitance. It stores and releases magnetic energy.

1.12   Equation tables

While cgs units are used in physics and in the study of magnetic materials, SI units are utilized when engineers investigate energy conversion in electric motors/ generators, as well as in electromagnetic wave propagation. Parameters are listed in Table 1.1. Note that two parameters, *magnetic susceptibility* and *permeability* of material, are included that have not been discussed previously. The concepts are very straightforward, and are used mainly in the field of electromagnetic wave study, frequently in SI units. They are listed at the bottom of the table.

Homework

Q1.1 A current I_0 passes through a straight, infinitely long thin film stripe conductor. The film thickness is t and the width is L. The current density in the conductor is uniform. Calculate the magnetic field H on the surface of the conductor. Do this based on Ampère's Law. See Fig. 1.Q1.

A1.1 Assuming that H is uniform on the top surface of the conducting thin film, Ampère's Law states that

$$H \cong I/(2L + 2t).$$

Since L $\gg$ t,

$$H \cong I/(2L)$$

in SI units, and

$$H \cong 4\pi I/(2L) = 2\pi I/L$$

in cgs units.

Q1.2 Same as in Q1.1. The film width is $L = 0.25\,\mu\text{m}$. The current density in the conductor is uniform. Calculate the magnetic field H as a function of the spacing from the top surface at the middle of the film $(x, y, z) = (0.125, 0, s)$. Do this for three film thicknesses $t = 0.1$, 0.2, 0.3 μm. See Fig. 1.Q2 for reference. Use Biot–Savart's Law and numerical integration.

A1.2 The current density J of the conductor is I/A, where A is the cross-sectional area. The current flowing through a small element of area dA is given by $dI = J\,dA$. The distance from this current element to the point of interest $(0.125, 0, s)$ is $r = \sqrt{(x - 0.125)^2 + y^2 + (z + s)^2}$, where s is the spacing.

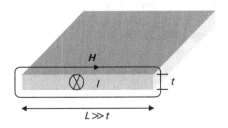

Figure 1.Q1. Current-induced magnetic field over a very large conducting film.

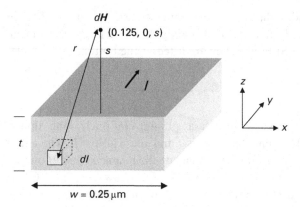

Figure 1.Q2. Magnetic field induced by a current in a thick conducting film of finite width.

Each current element contributes a field vector dH. Biot–Savat's Law states that

$$dH = \frac{1}{4\pi r^2}(J \cdot dAdy) \cdot r \times z$$

(in SI units). One can integrate the field across the cross-section of the conductor and then along the length of the conductor to obtain the field H for a total current across the conductor.

The result is shown in Chapter 5, Fig. 5.10. By normalizing the H with the current in the conductor, one obtains an important parameter: the flux conversion efficiency (H/I), with cgs units of Oe/mA. Clearly, a thicker conductor is less efficient than a thinner one. Similarly, the efficiency drops as the spacing of point of interest increases. The maximum flux is determined by the maximum current a conductor can carry, which is limited by the electromigration of the wire. It is an important parameter in the design of field-MRAM.

References

[1] C. Kittel, *Introduction to Solid State Physics*, 3rd edn (New York: John Wiley & Sons, Inc., 1968).
[2] N. Spaldin, *Magnetic Materials Fundamentals and Device Applications* (Cambridge: Cambridge University Press, 2003).
[3] W. E. Brown, Jr., *IEEE Trans. Magnetics* **20**(1), 112 (1984).
[4] R. E. Collin, *Field Theory of Guided Waves* (New York: McGraw-Hill, 1960), sect. 1.1.

2 Magnetic films

2.1 Origin of magnetization

The origins of magnetism are electron spin and its orbital motion around the atom nucleus. In this section we discuss the fundamental concept of magnetic moment with an individual atom and the origin of the electron's magnetic moment, which is a result of its angular momentum.

We can consider an electron moving in an orbital about an atomic nucleus. This would be equivalent to a current in cgs units given by

$$i = \frac{e}{c\tau},$$
(2.1)

where e is the charge of the electron, c is the velocity of light, e/c is the charge in emu and τ is the orbital period. From Eq. (1.9), this gives an orbital magnetic moment

$$\begin{aligned} \boldsymbol{m} &= i A \boldsymbol{n} \\ &= -\frac{eA}{c\tau}\boldsymbol{n}. \end{aligned}$$
(2.2)

The angular momentum of such an orbital can be expressed as follows:

$$\boldsymbol{p} = m_e r^2 \frac{d\theta}{dt}\boldsymbol{n},$$
(2.3)

where m_e is the mass of an electron and r is the radius of orbit. The area of orbit can be expressed as

$$A = \frac{1}{2}\frac{\tau p}{m_e}.$$
(2.4)

Therefore, we can write down the orbital magnetic moment of an electron in terms of the orbital angular momentum:

$$\boldsymbol{m} = -\frac{e}{2m_e c}\boldsymbol{p}.$$
(2.5)

The electron spin also contributes to the magnetic moment. The electronic spin angular momentum also generates a spin magnetic moment. The relation is given by

$$\boldsymbol{m}_S = -\frac{e\boldsymbol{p}_s}{m_e c},$$
(2.6)

where m_s is the spin magnetic moment and p_s is the electronic spin angular momentum. If we consider the total magnetic moment per electron as the vector sum of the orbital and spin magnetic moments,

$$m_t = m + m_S$$
$$= -\frac{e}{2m_ec}p - \frac{e}{2m_ec}2p_s, \tag{2.7}$$

where m_t is the total magnetic moment. The total magnetic moment can also be expressed as

$$m_t = -g\left(\frac{e}{2m_ec}\right)p_t, \tag{2.8}$$

where p_t is the total angular momentum of the electron. The factor g is called the Lande splitting factor, where $g = 1$ for orbital components only and $g = 2$ for spin components only of magnetic moment. The value of g must be between 1 and 2 depending on the relative sizes of the contributions from the orbit and spin to the total angular momentum.

We now describe the value of angular momentum in quantum terms. The possible values of angular momentum are restricted; therefore the magnetic moments are also quantized. Four quantum numbers are used to define uniquely each electron in an atom. These are the principal quantum number n, the angular momentum quantum number l, the orbital magnetic quantum number m_l, and spin projection quantum number m_{sq}.

The principal quantum number n is introduced by Bohr [1]. It determines the shell of the electron and has a dependence on the distance between the electron and the nucleus. The energy of an electron with principal quantum number is given by

$$E_n = -\frac{Z^2 m_e e^4}{8h^2 \varepsilon_0^2 n^2}, \tag{2.9}$$

where Z is the atomic number, h is Planck's constant and ε_0 is the permittivity of free space (in the Gaussian cgs system, $4\pi\varepsilon_0$ is replaced by 1). The principal quantum is a natural number ($n = 1, 2, 3\ldots$).

The angular momentum quantum number gives the orbital angular momentum through the following relation [2]:

$$P = l\left(\frac{h}{2\pi}\right). \tag{2.10}$$

This quantum number specifies the shape of an atomic orbital and strongly influences chemical bonds and bond angles. It can take values of $l = 0, 1, 2, 3,\ldots,(n-1)$, where n is the principal quantum number. In some contexts, $l = 0$ is called an s orbital, $l = 1$, a p orbital, $l = 2$, a d orbital and $l = 3$, an f orbital.

The orbital magnetic quantum number is the projection of the orbital angular momentum along a specified axis and determines the energy shift of an atomic orbital due to an external magnetic field. The component l_z of angular momentum l along the axis of a magnetic field is restricted to discrete values by quantum mechanics:

$$l_z = m_l \left(\frac{h}{2\pi} \right),$$ (2.11)

where the z-axis is defined as the axis perpendicular to the plane. The values of m_l are restricted to $m_l = -l, -l+1,\ldots, -1, 0, 1,\ldots, l-1, l$.

Electrons have spin angular momentum, which can be represented by the spin quantum number s. The value of spin quantum number is always $1/2$. Electron spins are constrained to lie either parallel or antiparallel to a magnetic field. The orientation of the electron can be represented by the spin projection quantum number m_{sp}, which is always constrained to have the values $\pm 1/2$.

If we now insert the angular momentum quantum number into Eq. (2.5), the magnetic moment arising from the orbital angular momentum can be expressed as follows:

$$m = \frac{eh}{4\pi m_e c} l.$$ (2.12)

This means that the magnetic moment contributed by the orbital angular momentum is an integral multiple of $eh/4\pi m_e c$. The quantity is known as the Bohr magneton, μ_B, which has a value of 9.27×10^{-21} erg/Oe. Therefore Eq. (2.12) can be rewritten as

$$m = \mu_B l.$$ (2.13)

The magnetic moment generated by the electronic spin angular momentum can be expressed as follows:

$$\begin{aligned} m_S &= 2 \frac{eh}{4\pi m_e c} s \\ &= 2\mu_B s. \end{aligned}$$ (2.14)

The value of spin quantum number is always $1/2$, and therefore the electronic spin angular momentum is also an integral multiple of μ_B. The total magnetic moment of an electron can be expressed in terms of multiples of the total angular momentum of the electron

$$\begin{aligned} m_t &= m + m_S \\ &= \mu_B l + 2\mu_B s \\ &= g\mu_B j, \end{aligned}$$ (2.15)

where j is the total angular momentum quantum number.

The coupling between the orbital angular momentum and the spin angular momentum gives the total atomic angular momentum j. This can be obtained in two ways: Russell–Saunders coupling and jj coupling.

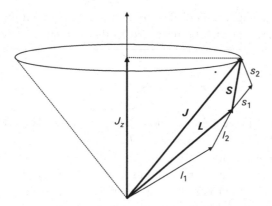

Figure 2.1. The total angular momentum with the vector addition of orbital momentum and spin momentum in Russell–Saunders coupling.

2.1.1   Russell–Saunders coupling

In light atoms, where spin-orbit interaction is weak, the coupling between the individual orbital angular momenta and the individual spins is stronger than the individual spin-orbit coupling. The total angular momentum is calculated by combining the orbital angular momenta of all the individual electrons (a vector sum) to obtain the total orbital momentum, and their spin angular momenta are added together to obtain the total spin momentum. The interaction between the spin and orbital angular momenta is called LS coupling or Russell–Saunders coupling.

The total spin momentum and the total orbital momentum are then combined to produce the total angular momentum (of all electrons of an atom):

$$J = \sum l_i + \sum s_i. \tag{2.16}$$

This leads to a total angular momentum J of the atom, which is simply the vector sum of the two non-interacting momenta L and S,

$$J = L + S, \tag{2.17}$$

as shown in Fig. 2.1.

2.1.2   jj coupling

In heavy atoms, generally $z>30$, where z is atomic number, the Russell–Saunders coupling scheme fails, because the spin and orbital angular momenta of individual electrons couple strongly. The coupling between spin and orbital angular momentum of each individual electron is much stronger than the coupling between different electrons. The resultant total angular momentum of each electron is:

$$j_i = l_i + s_i. \tag{2.18}$$

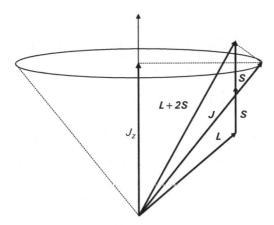

Figure 2.2. The total angular momentum with the vector addition of orbital momentum and spin momentum in jj coupling.

The resultant j_i then interact weakly via electrostatic coupling of their electron distributions to form a resultant total angular momentum:

$$J = \sum j_i = \sum (l_i + s_i), \tag{2.19}$$

as shown in Fig. 2.2.

2.2   Introduction of magnetic materials

In this section, we consider the characteristics of magnetic materials when a magnetic field passes through the material. The magnetic moments are arranged in different configurations in different types of materials under an applied field. The magnetic materials can be classified using some parameters. In Chapter 1, we mentioned that the relationship between B, H and M is given by

$$B = H + 4\pi M. \tag{2.20}$$

If the magnetic material is linear and isotropic , the magnetization is proportional to the applied field, and it can be written as follows:

$$M = \chi_m H, \tag{2.21}$$

where χ_m is magnetic susceptibility. Therefore, the relationship between magnetic induction B and magnetic field H can be written as

$$B = (1 + 4\pi\chi_m)H = \mu H, \tag{2.22}$$

where μ is the permeability,

$$\mu = 1 + 4\pi\chi_m. \tag{2.23}$$

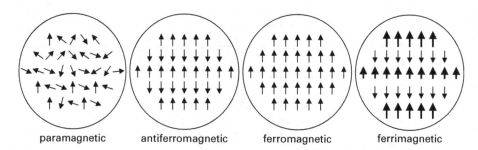

paramagnetic   antiferromagnetic   ferromagnetic   ferrimagnetic

Figure 2.3. Ordering of magnetic dipoles in different types of magnetic materials.

According to the permeability, magnetic materials can be roughly categorized into three types. If μ is slightly less than one (susceptibility slightly less than zero), the material is diamagnetic. If μ is slightly greater than one (susceptibility slightly greater than zero), the material is paramagnetic. If μ is much more than one (susceptibility much greater than zero), the material is ferromagnetic. However, there also exist some types of magnetic materials that are ferrimagnetic or antiferromagnetic. Figure 2.3 shows the different kinds of magnetic materials schematically. The reasons why there are different types of ordering and different resulting material properties comprise the subject of much of the rest of this chapter.

2.2.1   Diamagnetism

Langevin first figured out the theory of diamagnetism in 1905 [3]. He extended some of the earlier work of Ampère, Weber and Lenz. An electron moves around a nucleus in an atom in a way that is equivalent to current in a loop. Thus, the motion of the electron generates a magnetic moment, which can be written as

$$m = \frac{eA}{c\tau}n,\tag{2.24}$$

where n is in the direction normal to the electron orbital plane.

This assumes that the motion of the electron is circular, so the area A can be written as

$$A = r^2\pi,\tag{2.25}$$

where r is the radius of the orbital. The period τ can be written as

$$\tau = \frac{2r\pi}{v},\tag{2.26}$$

where v is the instantaneous tangential velocity of the electron. Thus, the relation between m and v is given by

$$m = \frac{evr}{2c}n.\tag{2.27}$$

The change of velocity will give rise to a change in magnetic moment. When a magnetic field is applied to the electron and the nucleus, the change in magnetic flux through the loop gives rise to an emf ξ in the loop, given by

$$\xi = -\frac{d(HA)}{dt}.$$

(2.28)

The electric field E acting around a circular loop can be written as

$$E = -\frac{A}{2r\pi}\frac{dH}{dt}.$$

(2.29)

The force exerted on the electron by the electric field is given by

$$F = \frac{e}{c}E = m_e\frac{dv}{dt}.$$

(2.30)

The change of velocity can be written as

$$\frac{dv}{dt} = -\frac{eA}{m_e cL}\frac{dH}{dt} = -\frac{er}{2m_e c}\frac{dH}{dt}.$$

(2.31)

On integrating over a change in magnetic field from zero to H, we obtain the change of velocity:

$$\int_{v_1}^{v_2} dv = -\frac{er}{2m_e c}\int_0^H dH,$$

(2.32)

$$v_2 - v_1 = -\frac{er}{2m_e c}H.$$

(2.33)

According to the change of velocity, the change of magnetic moment is given by

$$\Delta m = -\frac{e^2 r^2}{4m_e c^2}H.$$

(2.34)

This result applies only when the magnetic field is applied perpendicular to the plane of the loop. In general, the applied magnetic field is applied at the following angle between the plane and the projection of the orbit radius r on the plane normal to the field:

$$R = r\sin\theta.$$

(2.35)

Therefore, the change of magnetic moment will become

$$\Delta m = -\frac{e^2}{4m_e c}H\int_A^{R^2}\sin^2\theta\, dA,$$

(2.36)

where A is the area of a hemisphere. After integrating we get

$$\Delta m = -\frac{e^2 r^2}{6m_e c^2}H.$$

(2.37)

This is the case for a single electron. If we consider the moment for the case with Z electrons in the atom, the magnetic moment becomes

$$\Delta m_Z = -\frac{Ze^2r^2}{6m_ec^2}H. \tag{2.38}$$

Note that the number of atoms per unit volume is given by

$$n = \frac{N_0\rho}{m_a}, \tag{2.39}$$

where N_0 is Avogadro's number, ρ is the density, and m_a is the relative atomic mass. Therefore, the magnetization of the atom is given by

$$\Delta M = -\frac{N_0\rho}{m_a}\frac{Ze^2r^2}{6m_ec^2}H. \tag{2.40}$$

From Eq. (2.40), we can know that there is no net magnetic moment in the absence of a magnetic field in a diamagnet. The magnetic susceptibility is given by

$$\chi_m = \frac{M}{H} = -\frac{N_0\rho}{m_a}\frac{Ze^2r^2}{6m_ec^2}. \tag{2.41}$$

This shows us that the magnitude of magnetic susceptibility is always slightly less than zero.

2.2.2   Paramagnetism

The first measurement of magnetic susceptibility in paramagnetic material was made by Curie in 1895 [4]. He found that the magnetic susceptibility varied inversely with temperature:

$$\chi_m = \frac{C}{T}, \tag{2.42}$$

where C is the Curie constant. The first theory that explained Curie's measurement was proposed by Langevin in 1905 [3]. He assumed that the net magnetic moment of atoms and molecules is the same in paramagnetic materials. The magnetic moment is random in the absence of a magnetic field, and therefore the magnetization of the specimen is zero. When a magnetic field is applied, the magnetic moment tends towards the direction of the applied magnetic field. The energy of the moment in an applied field is given by

$$E = -mH\cos\theta. \tag{2.43}$$

Considering the thermal effect, the thermal energy tends to randomize the alignment of the moments. The probability p of an atom having an energy E is proportional to the Boltzmann factor:

$$p(E) = \exp(-E/k_BT), \tag{2.44}$$

where k_B is the Boltzmann constant. The probability function for the case of an isotropic material may be evaluated. The number of moments between θ and $\theta + d\theta$ is proportional to the surface area, now multiplied by the Boltzmann factor, i.e.

$$dn = A2\pi \exp(mH \cos \theta / k_B T) \sin \theta \, d\theta, \tag{2.45}$$

where A is a constant which gives the total number of moments per unit volume:

$$\int_0^n dn = n. \tag{2.46}$$

Integrating this expression over a hemisphere yields the total number of moments per unit volume:

$$2\pi A \int_0^\pi \exp\left(\frac{mH}{k_B T} \cos \theta\right) \sin \theta \, d\theta = n. \tag{2.47}$$

The magnetization is given by multiplying the number of atoms per unit volume and the component of magnetic moment then integrating over the total number:

$$
\begin{aligned}
M &= \int_0^n m \cos \theta \, dn \\
&= 2\pi A \int_0^\pi m \exp\left(\frac{mH}{k_B T} \cos \theta\right) \cos \theta \sin \theta \, d\theta.
\end{aligned} \tag{2.48}
$$

Substituting Eq. (2.47) into Eq. (2.48), we have

$$M = \frac{n \int_0^\pi m \exp\left(\frac{mH}{k_B T} \cos \theta\right) \cos \theta \sin \theta \, d\theta}{\int_0^\pi \exp\left(\frac{mH}{k_B T} \cos \theta\right) \sin \theta \, d\theta}. \tag{2.49}$$

To evaluate the integrals, we use the expression

$$
\begin{aligned}
M &= nm\left[\coth\left(\frac{mH}{k_B T}\right) - \frac{k_B T}{mH}\right] \\
&= nmL\left(\frac{mH}{k_B T}\right).
\end{aligned} \tag{2.50}
$$

This is the Langevin equation for a paramagnetic material; $L(x)$ is called the Langevin function. Figure 2.4 shows the Langevin function and expresses it as a series:

$$L(x) = \frac{x}{3} - \frac{x^3}{45} + \frac{2x^5}{945} - \cdots. \tag{2.51}$$

In most cases, $mH \ll k_B T$, so that the magnetization M becomes the first term in the series:

$$M = \frac{nm^2 H}{3k_B T}. \tag{2.52}$$

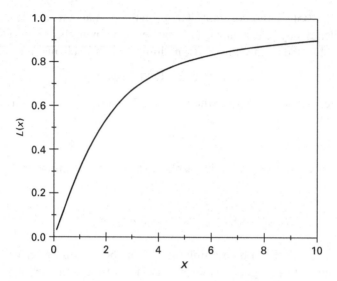

Figure 2.4. Dependence of the Langevin function $L(x)$ on x.

The Langevin theory leads to the Curie Law, since

$$\chi_m = \frac{M}{H} = \frac{nm^2}{3k_BT},$$ (2.53)

which is Curie's Law, with the Curie constant given by

$$C = \frac{nm^2}{3k_B}.$$ (2.54)

However, a lot of paramagnetic materials obey the more general law

$$\chi_m = \frac{C}{T - T_c}.$$ (2.55)

This is called the Curie–Weiss law [5]. Figure 2.5 shows the dependence of the magnetic susceptibility on temperature for both the Curie Law and the Curie–Weiss Law. Weiss explained that the individual atomic magnetic moments interact with each other via the molecular field, which is directly proportional to the magnetization [6]:

$$H_m = \gamma M,$$ (2.56)

where H_m is the molecular field and γ is the molecular field constant. The total magnetic field acting on the material then becomes

$$H_t = H + H_m = H + \gamma M.$$ (2.57)

The Curie Law can be rewritten as

$$\frac{C}{T} = \frac{M}{H + \gamma M}.$$ (2.58)

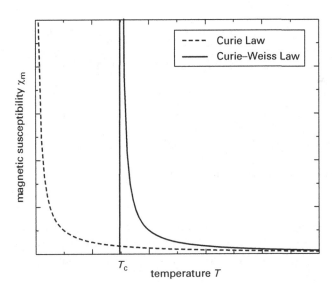

Figure 2.5. Dependence of the magnetic susceptibility on temperature for the Curie Law and the Curie–Weiss Law.

Therefore, the magnetic susceptibility becomes

$$\chi_m = \frac{C}{T - \gamma C} = \frac{C}{T - T_c}. \tag{2.59}$$

This is the Curie–Weiss Law and T_c is known as the Curie temperature.

2.2.3 Ferromagnetism

In ferromagnetic materials, the magnetic moments within domains are aligned parallel at temperatures below the Curie temperature. At higher temperatures (above the Curie temperature) ferromagnetic materials change to a paramagnetic state. For example, the Curie temperature of iron (Fe) is 770 °C, that of nickel (Ni) is 358 °C, and that of cobalt (Co) is 1131 °C. Figure 2.6 shows the alignment of magnetic moments in Fe, Ni and Co [7].

There was no theoretical understanding of ferromagnetism until Weiss proposed his hypothesis of the molecular field. Weiss supposed that any atomic magnetic moment $\boldsymbol{m}_i$ was affected by an effective field due to another moment $\boldsymbol{m}_j$. Assume that this field is in the direction of $\boldsymbol{m}_j$. It can be expressed as

$$\boldsymbol{H}_{ij} = \alpha_{ij}\boldsymbol{m}_j. \tag{2.60}$$

The total exchange interaction field at the moment $\boldsymbol{m}_i$ is summed over other moments:

$$\boldsymbol{H}_i = \sum_j \alpha_{ij}\boldsymbol{m}_j. \tag{2.61}$$

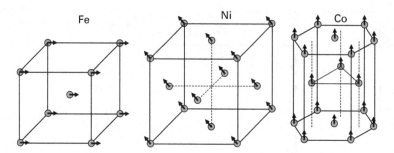

Figure 2.6. Alignment of magnetic moments in Fe, Ni and Co.

We can assume that the interactions between all moments are identical and the displacements between the moments are independent of one another; then, all of the α_{ij} are equal. Let γ substitute for α_{ij}. Therefore, the exchange interaction field can be written as follows:

$$H = \gamma M. \tag{2.62}$$

This is the original formulation of the Weiss Law. If we consider the case of the absence of an applied magnetic field, the field affected by the spontaneous magnetization is shown in Eq. (2.62). Substituting Eq. (2.62) into the Langevin equation, the spontaneous magnetization can be expressed as follows:

$$\frac{M}{M_0} = \coth\left(\frac{\gamma m M}{k_B T}\right) - \frac{k_B T}{\gamma m M}. \tag{2.63}$$

At a temperature corresponding to the Curie temperature the spontaneous magnetization tends to zero and thus the ferromagnetic state becomes to a paramagnetic state. If we consider the case with an applied magnetic field, Eq. (2.63) can be expressed as follows:

$$\frac{M}{M_0} = \coth\left[\frac{m(H_a + \gamma M)}{k_B T}\right] - \frac{k_B T}{m(H_a + \gamma M)}, \tag{2.64}$$

where H_a is the applied magnetic field. From the Curie–Weiss Law the Curie temperature takes the form

$$T_c = \frac{\gamma n m^2}{3 k_B}. \tag{2.65}$$

Heisenberg showed that the existence of a Weiss "molecular field" could be explained using a quantum mechanical treatment of the many-body problem. There is a term, of electrostatic origin, in the energy of interaction between neighboring atoms, which tends to orient the electron spins parallel to each other. This term is called the "exchange integral." It does not have a classical analog.

The magnitude of the Coulomb repulsion between two electrons at a distance of 1 Å is of the order of $10^5 J$, where J is the total angular momentum. A slight change

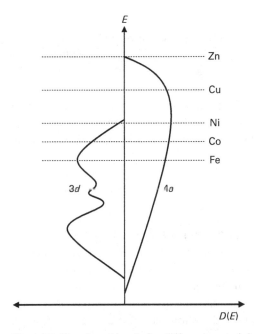

Figure 2.7. Density of levels for different materials in the $3d$ and $4s$ bands.

in the electron distribution is sufficient to create a huge field. It is 10^4 times larger than the magnetic dipolar interaction. Thus, if the electron distribution is changed, even by a small amount, the effect on the total energy of an atom can be significant. This explains why the molecular field is so large.

Band theory provides a good insight into the exchange field. In elemental transition metals, Fe, Ni and Co, the Fermi energy exists in a region of overlapping $3d$ and $4s$ bands. With rigid band approximation, we can picture the exchange interaction as shifting the energy of the $3d$ band for electrons with one spin direction relative to the band for electrons with the opposite spin direction. If the Fermi level is within the $3d$ band, then the displacement will lead to more electrons of the lower-energy spin direction and hence a spontaneous magnetic moment in the ground state. The shift in the $3d$ band is bigger than the shift in the $4s$ band (as shown in Fig. 2.7).

For the example of Ni, the exchange interaction displacement is so strong that the $3d$ sub-band is filled with five electrons and the other contains all 0.54 holes. So the saturation magnetization of Ni is $M_S = 0.54 N \mu_B$, where N is the total number of Ni atoms in the sample. We see why the magnetization of transition metals does not correspond to the number of electrons. In Cu, the Fermi level is in the $4s$ band, above the $3d$ band, which is totally filled. The $4s$ band has no exchange splitting, and then the number of up- and down-spin electrons are equal. Cu is not ferromagnetic.

Magnetocrystalline energy

The magnetization in ferromagnetic crystals tends to align along certain preferred crystallographic directions. The preferred directions are called the "easy" axes, since it is easiest to magnetize a demagnetized sample to saturation of the external field when the field is applied along a preferred direction.

Different materials have different easy axes; for example, in body-centered cubic (bcc) Fe, the easy axis is the $<100>$ direction (the cube edge). The body diagonal is the hard axis of the magnetization, and other orientations, such as the face diagonal, are intermediate.

By contrast, the easy axis of face-centered cubic (fcc) Ni is the $<111>$ body diagonal, and that of hexagonal close-packed (hcp) Co is the $<0001>$ direction.

Symmetry of magnetocrystalline anisotropy

The phenomenon that causes the magnetization to align itself along a preferred crystallographic direction is the magnetocrystalline anisotropy. The crystal is higher in energy when the magnetization points along the hard direction than when it points along the easy axis, and the energy difference between samples magnetized along easy and hard directions is called the magnetocrystalline anisotropy energy.

The symmetry of the magnetocrystalline anisotropy is always the same as that of the crystal structure. As a result, in iron, which is cubic, the anisotropy energy, E, can be written as a series expansion of the direction cosines, α_1, of the saturation magnetization relative to the crystal axes:

$$E = K_1 \left(\alpha_1^2 \alpha_2^2 + \alpha_2^2 \alpha_3^2 + \alpha_3^2 \alpha_1^2 \right) + K_2 \left(\alpha_1^2 \alpha_2^2 \alpha_3^2 \right) + \cdots \qquad (2.66)$$

Here K_1 and K_2 etc. are called the anisotropy constants. Typical values for iron at room temperature are $K_1 = 4.2 \times 10^5 \, \mathrm{erg/cm^3}$ and $K_2 = 1.5 \times 10^5 \, \mathrm{erg/cm^3}$. The energy is stored in the crystal when work is done against the anisotropy "force" to move the magnetization away from an easy direction. Note that the anisotropy energy is an even function of the direction cosines, and that it is invariant under interchange of the α_i among themselves.

Cobalt is hexagonal, with the easy axis along the hexagonal, c-axis. The anisotropy energy is uniaxial, and its angular dependence is a function only of the angle θ between the magnetization vector and the hexagonal axis. In this case, the anisotropy energy can be expanded as follows:

$$E = K_1 \sin^2 \theta + K_2 \sin^4 \theta + \cdots \qquad (2.67)$$

Typical values of the anisotropy constant for cobalt at room temperature are $K_1 = 4.1 \times 10^6 \, \mathrm{erg/cm^3}$ and $K_2 = 1.0 \times 10^6 \, \mathrm{erg/cm^3}$. Note that, in all materials, the anisotropy decreases with increasing temperature, and near T_c there is no preferred orientation.

Ferromagnetic thin film

A magnetic thin film is made of many small magnetic particles. The magnetic properties of small particles are dominated by the fact that, below a certain critical size, a particle contains only a single domain. The width of a domain wall depends on the balance between the exchange energy (which prefers a wide wall) and the magnetocrystalline anisotropy energy (which prefers a narrow wall). The balance results in typical domain wall widths of 100 nm. So, quantitatively, if a particle is smaller than 100 nm, a domain wall simply cannot fit inside it, resulting in a single-domain particle.

The magnetostatic energy and domain wall energy can be summed up and compared for a single-domain and two-domain sphere. The reduction in magnetostatic energy is proportional to the volume of the particle (i.e. r^3, where r is the particle radius), and the increase in the domain wall energy is proportional to the area of the wall, r^2. Thus, below a critical radius, r_c, it is energetically unfavorable to form domain walls, and a single-domain particle is formed.

Magnetic thin films can be viewed as a film containing a few layers of a single-domain particle. The particles interact with each other. The strength of interaction determines the domain properties of the thin film. In the middle of a large ferromagnetic film, the interparticle exchange energy aligns the magnetization of the particles. The magnetization can fluctuate around this direction under thermal excitation, and the time average of the magnetization remains in the same direction.

Magnetostriction

When a ferromagnetic material is magnetized it undergoes a change in length known as its magnetostriction. The origin of magnetostriction is the spin-orbit coupling, as it was for the crystalline anisotropy. Some materials, such as iron, elongate along the direction of magnetization and are said to have a positive magnetostriction. Others, such as nickel, contract and have negative magnetostriction. The length changes are very small – tens of parts per million, but nevertheless influence the easy axis when the film is under uniaxial stress.

For films with positive magnetostriction, the magnetostriction coefficient λ and uniaxial stress-induced anisotropy $H_{K\lambda}$ can be expressed as follows:

$$H_{K\lambda} \sim \lambda\sigma, \tag{2.68}$$

where σ denotes the uniaxial stress. The direction of $H_{K\lambda}$ depends on the sign of the magnetostriction. If $\lambda > 0$, then $H_{K\lambda}$ lies along the tensile stress. If $\lambda < 0$, $H_{K\lambda}$ lies perpendicular to the tensile stress. Under isotropic stress, there is no effect.

The vector sum of crystalline anisotropy and the stress-induced anisotropy are shown in Fig. 2.8. This resembles the usual vector sum, except that the angles of $H_{K\lambda}$ are multiplied by 2. For example, with $H_{K\lambda}$ at 90° and H_K at 0°, the vector sum is given by

$$H_{K\text{total}} = |H_K - H_{K\lambda}| \tag{2.69}$$

The vector sum points at 0° if $H_K > H_{K\lambda}$ and at 180° if $H_K < H_{K\lambda}$.

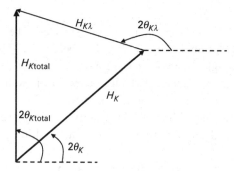

Figure 2.8. Vector addition of crystalline anisotropy and the stress-induced anisotropy.

Clearly, the easy axis of materials with large magnetostriction coefficient can be affected by stresses through material processing, either intentional or unintentional. By mixing Fe and Ni to form a $Ni_{79}Fe_{21}$ alloy, the magnetostriction can be cancelled out. This alloy, of great importance in recording head technology, is called Permalloy.

2.2.4   Antiferromagnetism

In antiferromagnetic materials the nearest-neighbor moments are aligned antiparallel. The theory of antiferromagnetism was chiefly developed by Néel [8]. We can imagine that the sublattices can be divided into two types, A and B, in an antiferromagnetic material and that sublattice A has only sublattice B as nearest neighbors. The interaction between the moments on their own sublattice yields a positive coupling coefficient, whereas the interaction between moments on different sublattices yields a negative coupling coefficient.

Figure 2.9 shows the dependence on temperature of the inverse of the magnetic susceptibility in an antiferromagnetic material. As the temperature decreases, the magnetic susceptibility increases until it reaches a maximum at a critical temperature, T_N, called the Néel temperature. When the temperature is above T_N the material is in a paramagnetic state and it is in an antiferromagnetic state at below T_N. The antiferromagnetic material still obeys the Curie–Weiss Law

$$\chi_m = \frac{C}{T - \theta},\tag{2.70}$$

but with a negative value of θ.

In order to understand this in detail, the molecular field theory is applied. Assume that only the interaction between nearest neighbors is considered. We now have two molecular fields to deal with. One is molecular field $\boldsymbol{H}_{mA}$, which is acting on sublattice A. The other one is molecular field $\boldsymbol{H}_{mB}$, which is acting on sublattice B:

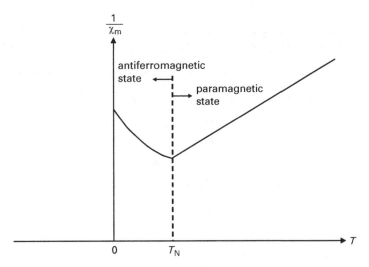

Figure 2.9. Dependence of the inverse of the magnetic susceptibility on temperature in an antiferromagnetic material.

$$H_{mA} = -\gamma M_B, \tag{2.71}$$

$$H_{mB} = -\gamma M_A, \tag{2.72}$$

where γ is the molecular field constant, M_B is the spontaneous magnetization of sublattice B and M_A is the spontaneous magnetization of sublattice A. At temperatures below T_N, sublattices A and B are spontaneously magnetized and

$$M_A = -M_B. \tag{2.73}$$

At temperatures above T_N, the material is in a paramagnetic state. The Curie Law under an applied magnetic field can be written as

$$\frac{M_A}{H - \gamma M_B} = \frac{C'}{T}, \tag{2.74}$$

$$\frac{M_B}{H - \gamma M_A} = \frac{C'}{T}, \tag{2.75}$$

where C' is the Curie constant of each sublattice and H is the applied magnetic field. We rewrite Eqs. (2.74) and (2.75) as follows:

$$M_A T = C'(H - \gamma M_B), \tag{2.76}$$

$$M_B T = C'(H - \gamma M_A). \tag{2.77}$$

By adding Eqs. (2.76) and (2.77), we can find the total magnetization produced by the magnetic field:

$$(M_A + M_B)T = 2C'H - C'\gamma(M_A + M_B), \tag{2.78}$$

$$MT = 2C'H - C'\gamma M. \tag{2.79}$$

Table 2.1 Néel temperature of some antiferromagnetic materials

Antiferromagnetic materials	*MnO*	*NiO*	*CoO*	*FeO*	*Fe$_2$O$_3$*	*MnF$_2$*	*FeF$_2$*	*MnO$_2$*	*NiF$_2$*
Néel temperature T_N (K)	122	523	293	198	950	950	79	84	78

Therefore, the magnetic susceptibility is given by

$$\chi_m = \frac{M}{H} = \frac{2C'}{T + C'\gamma}.$$

(2.80)

This relation is equivalent to Eq. (2.80), with

$$C = 2C',$$

(2.81)

$$\theta = C'\gamma.$$

(2.82)

Now we consider the condition in the absence of an applied magnetic field at temperature T_N. Equation (2.77) becomes

$$M_A T_N = -C'\gamma M_B.$$

(2.83)

Therefore, the Néel temperature is given by

$$T_N = -C'\gamma \frac{M_B}{M_A} = -C'\gamma = -\theta.$$

(2.84)

Table 2.1 shows the Néel temperature of antiferromagnetic materials.

2.2.5   Ferrimagnetism

Ferrimagnetic materials comprise a particular case of antiferromagnetic materials. In ferrimagnetic materials the magnetic moments on sublattices A and B still align in the opposite direction, but the magnetic moments have different magnitudes. As in ferromagnetic materials, the spontaneous magnetization in ferrimagnetic materials disappears above a certain critical temperature, also called the Curie temperature, and the material changes to a paramagnetic state. Ferrimagnetic order was first suggested by Néel to expain the behavior of ferrites in 1948, and the word ferrimagnetism is due to him.

The general formula of a ferrimagnet is $MO \cdot Fe_2O_3$, where M is a transition metal, such as Mn, Ni, Fe, Co, Mg, etc. These ferrites are cubic, and the most familiar, Fe_3O_4, which may be called iron ferrite, is the oldest magnetic material.

Other ferrites are made up of the hexagonal ferrites such as $BaO \cdot 6(Fe_2O_3)$ and $SrO \cdot 6(Fe_2O_3)$. These are magnetically hard and have been used as permanent-magnet materials. Their Curie temperature is typically in the range 500~800 °C.

2.3   Ferromagnet/antiferromagnet bilayer structure

Meiklejohn and Bean first discovered the unidirectional anisotropy in 1956 when they were studying fine particles of cobalt with the cobaltous oxide shell [9, 10].

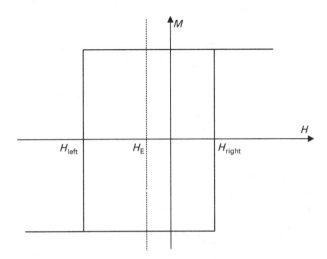

Figure 2.10. Schematic for H_{right}, H_{left} and H_E in a ferromagnet/antiferromagnet bilayer structure.

This anisotropy is the consequence of the interaction between a ferromagnet and an antiferromagnet, and it is best delineated as the exchange anisotropy. The exchange anisotropy creates only one easy direction of magnetization. One of the important phenomena associated with the exchange anisotropy is the hysteresis loop displacement. The displacement is quoted as the exchange bias, H_E, which can be defined as follows:

$$H_E = \frac{H_{left} + H_{right}}{2}, \qquad (2.85)$$

where H_{left} and H_{right} are the left and right external fields at the zero crossing of magnetization, respectively. A schematic diagram for H_E is shown in Fig. 2.10. Note that H_E is observed only if an external magnetic field is applied while the ferromagnetic/antiferromagnetic (FM/AFM) material is cooled through the Néel temperature of the antiferromagnet. Since the T_N of cobaltous oxide is 293 K, the cobalt with a cobaltous oxide shell was cooled from 300 K to 77 K in an external field measuring 1×10^4 Oe. Meiklejohn and Bean observed H_E to be -6×10^3 Oe at 77 K temperature, and it is a negative exchange bias (shown in Fig. 2.11 [9]).

In recent years, the exchange bias in thin films has found technological applications such as magnetic recording media, magnetic sensors or as stabilizers in magnetic reading heads, and MRAM. Research on the phenomenon of the exchange bias has attracted a lot of attention both in the theories and experiments.

2.3.1 Intuitive picture in exchange bias

The hysteresis loop of the ferromagnetic material exhibits symmetry about the origin of the coordinates. There is no H_E for the FM/AFM bilayer at temperatures larger than the "blocking temperature" (T_B). In general, the blocking temperature

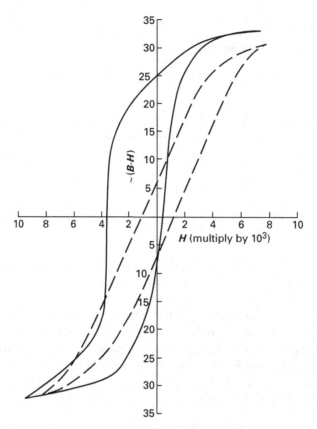

Figure 2.11. Measured hysteresis loops of cobalt with cobaltous oxide shell, after cooling from 300 K to 77 K, with applied field (solid loop) and without field (dashed loop) [9].

is smaller than the Néel temperature. The exchange bias can be qualitatively understood by assuming there to be an exchange interaction at an FM/AFM interface, since the spin orders disappear as $T > T_B$ [11].

When cooling (with an applied magnetic field) to temperature $T_N < T < T_C$ in an FM/AFM bilayer structure, the spins in a ferromagnet align toward the direction of the applied field. However, the spins in an antiferromagnet remain in random orientation (see Fig. 2.12(1)) because the antiferromagnet is in the paramagnetic state in that temperature range. When cooling to temperature T ($T < T_B < T_N$)[12], the antiferromagnet is in the antiferromagnetic state. Owing to the interaction at the FM/AFM interface, the spins at the antiferromagnetic interface orient ferromagnetically to those of the ferromagnetic spins (assuming a ferromagnetic interaction). For antiferromagnetic ordering within the antiferromagnet, the other spin plans order antiparallel (see Fig. 2.12(2)).

When the field is in the opposite direction, the spins start to rotate in a ferromagnet. However, for a large enough antiferromagnetic anisotropy, the spins remain unchanged in an antiferromagnet. The spins at the antiferromagnetic

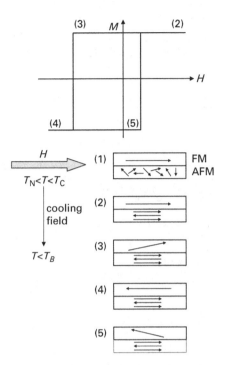

Figure 2.12. The intuitive picture in exchange bias at five different stages of the FM/AFM bilayer structure.

interface keep the ferromagnetic spins oriented in their original direction due to the interfacial exchange interaction between the FM and the AFM (see Fig. 2.12(3)). The spins have one single stable state in a ferromagnet, and the anisotropy is unidirectional. Thus, a larger applied field is required to rotate completely the spins in the reverse direction (see Fig. 2.12(4)). The applied field then returns to its original direction. The interfacial interaction between the FM and the AFM assists the ferromagnetic spins to become oriented in the direction of the applied field. The spins start to rotate with the application of a small applied field in a ferromagnet (see Fig. 2.12(5)). The material behaves as though there were an additional applied field. The hysteresis loop is shifted, as there is an extra magnetic field.

2.3.2 Positive exchange bias

A negative value of H_E is measured in most of FM/AFM bilayer structures, i.e. the hysteresis loop is shifted in the opposite direction from the cooling field. In 1996, Nogués, Lederman, Moran and Schuller studied the FM/AFM (Fe/FeF$_2$) bilayer structure and discovered a positive exchange bias [13].

The Fe^{2+} ions in FeF$_2$ form a body-centered tetragonal (bct) crystal structure, and FeF$_2$ has a large uniaxial magnetic anisotropy along the c-axis. Because the T_N of FeF$_2$ is 79 K, the Fe/FeF$_2$ interface was cooled from 120 K to 10 K in the

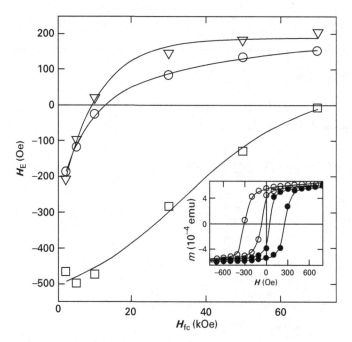

Figure 2.13. The H_E dependence of H_{fc} at $T = 10\,\text{K}$ for three different growth Fe/FeF$_2$ temperatures. □ grown at 200 °C; ▽ grown at 250 °C; ○ grown at 300 °C. Inset: The hysteresis loop of Fe/FeF$_2$ grown at 300 °C for ● is $H_{fc} = 70\,\text{kOe}$ and for ○ is $H_{fc} = 2$ kOe [13].

varied cooling field. The experimental results show that the hysteresis loops shift to the negative field in low cooling fields, which mirrors the normal behavior of other systems. Nevertheless, Nogués *et al.* observed a positive exchange bias ($H_E > 0$) in large cooling fields (see Fig. 2.13 [13]). Figure 2.13 shows the dependence of H_E on the cooling fields (H_{fc}) in Fe/FeF$_2$ at 10 K. The H_E becomes less negative with the increasing of H_{fc}. Furthermore, it may be observed that the positive H_E in Fe/FeF$_2$ increases at 300 and 250 °C. When cooled at 7×10^4 Oe, the magnitude of H_E is as large as the negative H_E observed in 2×10^3 Oe (see Fig. 2.13, inset).

The positive exchange bias can be explained by considering that the exchange interaction between the FM layer and the AFM layer competes with the coupling between the cooling field and the exchange field from the AFM interface, and the FM/AFM interfacial interaction must be antiferromagnetic. If the FM/AFM interfacial interaction is antiferromagnetic, and the cooling field is large enough to overcome the antiferromagnetic interaction, the spins in the AFM interface line up in the direction of the cooling field. Once the applied field weakens, the magnetization of the FM rotates from a positive direction to a negative direction of the applied field in order to reduce the interface exchange energy. In this case, the hysteresis loop shifts to a positive direction of the cooling field.

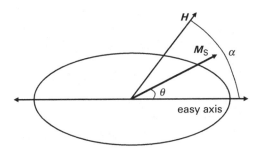

Figure 2.14. Schematic of ferromagnetic particle with anisotropy and applied field in the Stoner–Wohlfarth model.

2.3.3 Theories of exchange bias

Since the exchange bias was first observed in 1956, there has been a great deal of experimental and theoretical research undertaken on this amazing phenomenon. Even though many different theoretical models were proposed, a full understanding of the exchange bias still does not exist in theory. In Section 2.3 (particularly Section 2.3.2), we reviewed some experimental results of the exchange bias. In this section we review the important theoretical models, which help us to understand the mechanism of the exchange bias.

In the intuitive model, the magnetization of the FM layer and the spins of the AFM layer are assumed to exhibit coherent rotation. The Stoner–Wohlfarth model is used to formulate the hysteresis loop curve. In the Stoner–Wohlfarth model [14, 15], the energy density of the single-domain particle is written as follows:

$$E = K \sin^2 \theta - H M_S \cos(\alpha - \theta), \tag{2.86}$$

where K is the anisotropy of the particle, H is the applied field, M_S is the saturation magnetization of the particle, θ is the angle between M_S and the easy axis and α is the angle between the applied field and the easy axis (see Fig. 2.14). For simplicity, we assume the direction of the applied is parallel to the easy axis, and the equilibrium of M_S is given by

$$\frac{\partial E}{\partial \theta} = 2K \sin \theta \cos \theta + H M_S \sin \theta = 0; \tag{2.87}$$

$$\frac{\partial^2 E}{\partial \theta^2} = 2K \cos^2 \theta - 2K \sin^2 \theta + H M_S \cos \theta > 0, \tag{2.88}$$

and we define $M \equiv M_S \cos \theta$, $M/M_S \equiv m$, $H_K \equiv 2K/M_S$ and $H/H_K \equiv h$. Therefore, we have

$$(m + h)\sqrt{1 - m^2} = 0; \tag{2.89}$$

$$2m^2 + hm - 1 > 0. \tag{2.90}$$

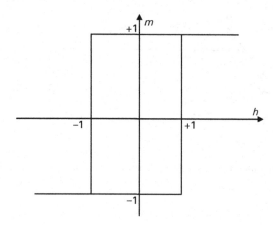

Figure 2.15. The hysteresis loop of a ferromagnetic particle with the direction of applied field parallel to the easy axis ($\alpha = 0$) in the Stoner–Wohlfarth model.

The hysteresis loop of a single-domain particle can be obtained as shown in Fig. 2.15. In the FM/AFM bilayer structure, the energy per unit interface area can be written as [16]

$$\varepsilon = -HM_{FM}t_{FM}\cos(\theta - \alpha) + K_{FM}t_{FM}\sin^2\alpha$$
$$+ K_{AFM}t_{AFM}\sin^2\beta + J_{EX}\cos(\alpha - \beta), \tag{2.91}$$

where H is the applied field, M_{FM} is the FM saturation magnetization, t_{FM} is the thickness of the FM layer, K_{FM} is the anisotropy of the FM layer, K_{AFM} is the anisotropy of the AFM layer, t_{AFM} is the thickness of the AFM layer and J_{EX} is the interfacial exchange integral. The angles are defined as follows: α is the angle between M_{FM} and the FM anisotropy axis, β is the angle between the AFM sublattice magnetization (M_{AFM}) and the AFM anisotropy and θ is the angle between the applied field and the FM anisotropy axis (see Fig. 2.16).

In general, K_{FM} is smaller than K_{AFM}. Neglecting the FM anisotropy, and minimizing with respect to α and β, the hysteresis loop shift obtained is

$$H_E = \frac{J_{EX}}{M_{FM}t_{FM}}. \tag{2.92}$$

The H_E obtained by the intuitive model takes values orders of magnitude larger than experimental results.

2.3.4   AFM domain wall model

To explain the discrepancy between the value of the exchange bias obtained using the intuitive model and experimental observations, Mauri, Siegmann, Bagus and Kay proposed a model that would reduce the interfacial exchange energy between

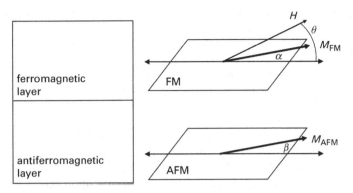

Figure 2.16. Schematic of the magnetic moments and the angles between the magnetic moments and the easy axis in the FM/AFM bilayer structure.

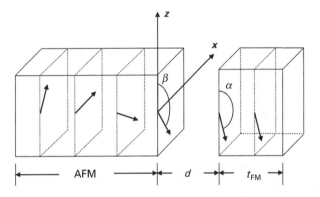

Figure 2.17. Schematic of the domain wall in the AFM of the FM/AFM bilayer structure [16].

the FM and AFM layer [17]. The main assumptions made are as follows: (a) the FM interface coupling is across a perfect flat interface; (b) an AFM domain can form at the interface; (c) the AFM layer is infinitely thick; (d) the spins within the FM rotate coherently. The AFM domain wall model is illustrated in Fig. 2.17.

Figure 2.17 depicts a situation in which an applied field is applied opposite to the z-axis and the exchange coupling occurs across the interface with thickness d. Therefore, the total magnetic energy per unit area is given by [17]

$$\varepsilon = 2\sqrt{A_{AFM}K_{AFM}}(1 - \cos \beta) + J_{EX}[1 - \cos(\beta - \alpha)] + K_{FM}t_{FM} \cos^2\alpha + HM_{FM}t_{FM}(1 - \cos \alpha),$$

(2.93)

where A_{AFM} is the exchange stiffness in the AFM layer. The first term is the energy of the tail of a domain wall extending into the AFM layer; the second term is the stiffness at the interface; the third term is the anisotropy energy in the FM layer; and the last term is the magnetostatic energy. The hysteresis loop curve is

calculated by minimizing the energy per unit area with α and β. The numerical calculation yields the following two limiting results:
strong interfacial coupling

$$H_E = -2\left(\frac{\sqrt{A_{AFM}K_{AFM}}}{M_S t_{FM}}\right) \text{ for } \sqrt{A_{AFM}K_{AFM}} \ll J_{EX}. \tag{2.94}$$

weak interfacial coupling

$$H_E = -\left(\frac{J_{EX}}{M_S t_{FM}}\right) \text{ for } \sqrt{A_{AFM}K_{AFM}} \gg J_{EX}. \tag{2.95}$$

In the strong interfacial coupling case, the H_E is limited at an energy far less than the fully uncompensated interfacial exchange coupling. In the weak interfacial coupling case, the H_E saturates at the strength of the interfacial exchange coupling. The AFM domain wall model reduces successfully the value of H_E, which is obtained by the intuitive model, and it highlights the formation of an AFM domain wall in the limit of strong interfacial exchange energy.

2.3.5   Random field model

Malozemoff proposed a random field model to explain the value of the exchange bias in 1987 [18]. He rejected the assumption of the perfect umcompensated AFM interface, and proposed a model of exchange anisotropy based on the assumption of an interfacial AFM moment imbalance that originated from features such as roughness and structural defects (see Fig. 2.18).

Therefore, for interfacial roughness that is random on an atomic scale, the local interfacial energy, σ_i, is also random, i.e.

$$\sigma_i = \pm z \frac{U}{a^2}, \tag{2.96}$$

where U is the interfacial exchange parameter, a is the cubic lattice constant and z is a number of order unity. The random field model argues that a net average interfacial energy will exist. The average σ in a region of area X^2 will go down statistically as

$$\sigma \sim \frac{\sigma_i}{\sqrt{X^2/a^2}}. \tag{2.97}$$

Although obtaining a domain of size X would lower the random field energy, K_{AFM} will limit the domain size. The anisotropy energy confines the domain width, and creates an additional surface energy of the domain wall. The balance between the exchange and anisotropy energy is given by

$$L \approx \pi \sqrt{A_{AFM}/K_{AFM}}. \tag{2.98}$$

Thus, the average interfacial exchange energy is given by

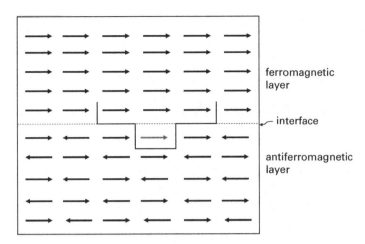

Figure 2.18. Schematic side view of spin configuration in the FM/AFM bilayer structure with AFM rough interface.

$$\Delta\sigma = \frac{4zU}{aL\pi}.$$ (2.99)

Accordingly, the exchange bias due to the random field model is obtained as

$$H_E = \frac{2z\sqrt{A_{AFM}K_{AFM}}}{M_{FM}t_{FM}\pi^2}.$$ (2.100)

The result is similar to the strong interfacial exchange coupling by the AFM domain wall model. Thus, the value of the H_E obtained by the random field model agrees with experimental results.

2.4   Interlayer exchange coupling in ferromagnet/metal/ferromagnet multilayer

For two ferromagnetic films that are separated by a non-magnetic metal (M) spacer layer, the magnetizations of two ferromagnetic layers couple to each other by an indirect exchange coupling through the electrons of the spacer layer. The indirect exchange coupling in the ferromagnet/metal/ferromagnet (FM/M/FM) multilayer is known as interlayer exchange coupling. The interlayer exchange coupling exhibits oscillatory behavior with respect to the thickness of the metal spacer, and it has been observed experimentally in many magnetic multilayer systems. Grünberg, Schreiber, Pang, Brodsky and Sowers first investigated exchange coupling in transition metals (Au and Cr) in 1986 [19]. However, the oscillatory exchange coupling was not observed in the rare-earth multilayers. In 1990, Parkin, More and Roche proposed the systematic study of the interlayer exchange coupling in some transition metal multilayers, and they found that the coupling oscillates as a function of the thickness of the spacer layer [20]. They show that in some systems (Co/Ru, Co/Cr, Fe/Cr) the magnitude of the

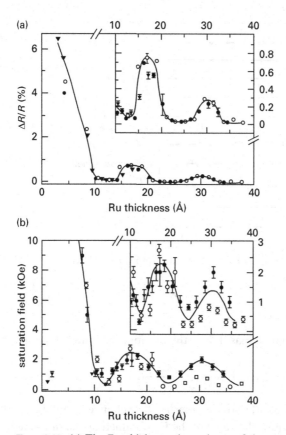

Figure 2.19. (a) The Ru thickness dependence of the saturation magnetoresistance at 4.5 K and (b) the Ru thickness dependence of the saturation field at 300 K in $[Co(20 \text{ Å})/Ru(t_{Ru})]_{20}$ deposited at temperatures of ●, 40 °C; ○, 125 °C [20].

saturation fields oscillates with the same period as that of the saturation magnetoresistance (MR). Figures 2.19 and 2.20 show the dependence of the saturation magnetoresistance and the saturation field on the thickness of the Ru and Cr layers [20]. The oscillatory behavior of the saturation magnetoresistance and saturation field indicates that the interlayer exchange coupling is oscillatory and depends on the metal thickness in the FM/M/FM multilayers.

Shortly thereafter, Parkin, Bhadra and Roche presented evidence for an oscillatory interlayer exchange coupling in one of the most nearly free-electron-like transition metals, Cu [21]. The direct evidence is that there appear giant oscillations in the saturation magnetoresistance of Co/Cu/Co multilayer structures as the Cu layer thickness is varied (see Fig. 2.21 [22]).

Figure 2.21 shows that the saturation magnetoresistance oscillates with the thickness of the Cu layer at 300 K and 4.2 K. In each case the period of the oscillation is about 10 Å. It should be remarked that the antiferromagnetic interlayer exchange coupling in Co/Cu is about 30 times smaller than the largest effect in Co/Ru at the peak of the first antiferromagnetic oscillation.

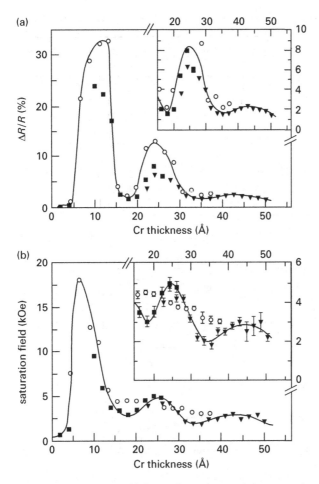

Figure 2.20. (a) The Cr thickness dependence of the saturation magnetoresistance at 4.5 K and (b) the Cr thickness dependence of the saturation field at 4.5 K in [Fe(20 Å)/Cr $(t_{Cr})]_N$ deposited at temperatures of ▼, ■, 40 °C ($N = 30$); ○, 125 °C ($N = 20$) [20].

Since the discovery of oscillatory interlayer exchange coupling between two ferromagnetic layers separated by thin layers of the non-magnetic metals the investigation of the FM/M/FM multilayer has attracted a great deal of interest [23–27]. Some of the experimental results are listed in Table 2.2 [27].

The J_1 referred to in the table is the estimated maximum strength of the interlayer exchange coupling as the thickness of the metal in the FM/M/FM multilayer is varied. We will discuss this from a theoretic viewpoint in Section 2.4.1.

2.4.1   Ruderman–Kittel–Kasuya–Yosida interaction

The RKKY interaction is the indirect interaction of two localized spins via a hyperfine interaction within the conduction electron sea [27, 28]. The localized

Table 2.2 Experimental results of observed coupling strengths and periods [27]

FM/M/FM	J_1 (mJ/m^2) at (thickness) (nm)	Periods in ML (nm)
Co/Cu/Co (100)	0.4 (1.2)	2.6 (0.47), 8 (1.45)
Co/Cu/Co (110)	0.7 (0.85)	9.8 (1.25)
Co/Cu/Co (111)	1.1 (0.85)	5.5 (1.15)
Fe/Au/Fe (100)	0.85 (0.82)	2.5 (0.51), 8.6 (1.75)
Fe/Cr/Fe (100)	1.5 (1.3)	2.1 (0.3), 12 (1.73)
Fe/Mn/Fe (100)	0.14 (1.32)	2 (0.33)
Co/Rh/Co (111)	34 (0.48)	2.7 (0.6)
Co/Ir/Co (111)	2.05 (0.5)	4.5 (1.0)
Co/Ru/Co (001)	6 (0.6)	5.1 (1.1)

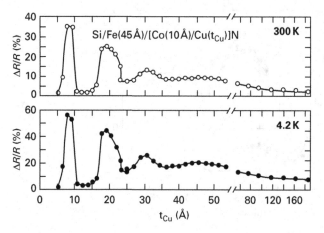

Figure 2.21. The Cu thickness dependence of the saturation magnetoresistance at 300 K and 4.2 K in [Co(10 Å)/Cu(t_{Cu})]$_N$ for $N = 16$ (● and ○), $N = 8$ (■ and □) [21].

spin ($\mathbf{I}_n$) in a metal senses the other localized spin ($\mathbf{I}_m$) direction in the following way. The contact interaction ($\mathbf{I}_n \cdot \mathbf{S}$) of the hyperfine coupling scatters a conduction electron having a given state of the electron spin S. The other localized spin $\mathbf{I}_m$ feels the density of the scattered electron through the interaction ($\mathbf{I}_m \cdot \mathbf{S}$), and thereby senses the localized spin $\mathbf{I}_n$. The two localized spins are coupled due to this process.

The RKKY interaction is given by [28]

$$E_{RKKY} = \frac{4J^2 m^* k_F^4}{(2\pi)^3} \times \frac{2k_F r \cos 2k_F r - \sin 2k_F r}{(2k_F r)^4} \mathbf{I}_n \cdot \mathbf{I}_m, \qquad (2.101)$$

where r is the distance between two localized spins, $\mathbf{I}_m$ and $\mathbf{I}_n$, and k_f is the Fermi wave vector. The indirect exchange coupling between two localized spins via the

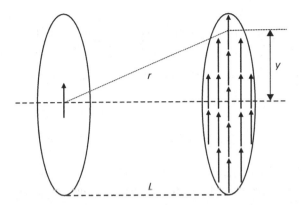

Figure 2.22. Two circular ferromagnetic layers separated by a distance L. The spins in the two ferromagnetic layers are aligned in the same direction.

conduction electron sea is as shown in Eq. (2.101), and the interaction oscillates with the distance of the two localized spins. It can be found that, not only the strength of the interaction is oscillatory, but also the sign of the interaction oscillates. This is one of the most important characteristics of the RKKY interaction.

After the oscillatory interlayer exchange coupling was observed in FM/M/FM structures, the first theoretical model was constructed and used to explain the experimental results that had been based on the RKKY interaction. It is instructive to calculate it using the free electron approximation. The model corresponds to a condition in which the electron sea fills the whole space. The magnetizations in the two ferromagnetic layers couple to each other due to the electrons in the non-magnetic metal, which is positioned between the two ferromagnetic layers. We assume that the two ferromagnetic layers are separated by distance L, and that the spins in the ferromagnetic layers are aligned in the same direction (see Fig. 2.22). The indirect exchange coupling between the two localized spins is as in Eq. (2.101). In order to obtain the interlayer exchange coupling between the two ferromagnetic disks, a sum over all the spins in the right-hand disk has to be performed. First, we calculate that one localized spin ($\mathbf{I}_l$) at the center of the left-hand ferromagnetic disk couples with all spins in the right-hand ferromagnetic disk. We approximate the summation by integration. The exchange energy is as follows:

$$E_1 = \frac{4J^2 m^* k_{\mathrm{F}}^4}{(2\pi)^3} \mathbf{I}_1 \cdot \mathbf{w}_\mathrm{r} \int_0^{2\pi} d\theta \int_0^a \frac{2k_{\mathrm{F}} r \cos 2k_{\mathrm{F}} r - \sin 2k_{\mathrm{F}} r}{(2k_{\mathrm{F}} r)^4} y \, dy, \qquad (2.102)$$

where $\mathbf{w}_\mathrm{r}$ denotes the spins per unit area in the right-hand ferromagnetic disk, a is the radius of the right-hand ferromagnetic disk and $r^2 = L^2 + y^2$. The integral result yields

$$E_1 = \frac{J^2 m^* k_F^2}{8\pi^2} \mathbf{I}_1 \cdot \mathbf{w}_r \left[-\int_l^{\tilde{l}} \frac{\sin t}{t}\, dt + \frac{\sin \tilde{l} - \tilde{l} \cos \tilde{l}}{\tilde{l}^2} + \frac{l \cos l - \sin l}{l^2} \right], \qquad (2.103)$$

where

$$l \equiv 2k_F L; \qquad (2.104)$$

$$\tilde{l} \equiv 2k_F \sqrt{L^2 + a^2}. \qquad (2.105)$$

Then we take $a \rightarrow \infty$, and change $\mathbf{I}_1$ to $\mathbf{w}_l$. This yields the interlayer exchange coupling between two ferromagnetic surfaces. The exchange energy per unit area between two ferromagnetic disks is given by

$$\varepsilon_A = \frac{J^2 m^* k_F^2}{8\pi^2} \mathbf{w}_l \cdot \mathbf{w}_r F_1(l), \qquad (2.106)$$

where

$$F_1(l) = \frac{l \cos l - \sin l}{l^2} - \int_l^\infty \frac{\sin t}{t}\, dt; \qquad (2.107)$$

$$\int_l^\infty \frac{\sin t}{t}\, dt = \frac{\pi}{2} - \int_0^l \frac{\sin t}{t}\, dt. \qquad (2.108)$$

The interlayer exchange coupling between two ferromagnetic disks separated by distance L exhibits the same behavior as $F_1(L)$ (shown in Fig. 2.23). We find that the strength of the indirect exchange coupling between two localized spins decays as r^{-3} (see Eq. (2.101)). However, the strength of the interlayer exchange coupling between two ferromagnetic disks depends on the inverse of the metal thickness, L^{-1} (see Eq. (2.103)).

2.4.2   Néel coupling

When the surface of the ferromagnetic layer is rough, there are magnetic poles at the surface (see Fig. 2.24). Therefore the fringe fields will generate from those magnetic poles. The fringe field is parallel to the magnetization of the ferromagnetic layer. On the other hand, if there is interfacial roughness in the FM/M/FM multilayer, the Néel coupling favors a parallel alignment between two ferromagnetic layers. The Néel coupling comprises the contribution from the magnetostatic interaction and was originally described by Néel in 1962 [29, 30].

The Néel coupling energy density of the two ferromagnetic layers separated by a non-magnetic spacer with thickness t which has a two-dimensional sinusoidal behavior with an amplitude h and a wavelength w is written as

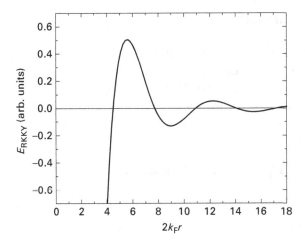

Figure 2.23. The F_1 function plotted against L, the distance between two ferromagnetic disks ($k_F = 1\,\text{Å}^{-1}$).

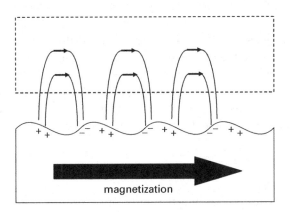

magnetization

Figure 2.24. Schematic figure of the Néel coupling from the interfacial roughness of the ferromagnetic layer.

$$J = \frac{\pi^2 h^2}{\sqrt{2}w}(\mu_0 M_1 M_2)e^{-(2\sqrt{2}\pi t)/w}. \tag{2.109}$$

It assumes that the magnetization of one ferromagnetic layer is fixed and the other free magnetization is affected by a magnetic field

$$H_N = \frac{J}{\mu_0 M_2 t_{\text{free}}} \tag{2.110}$$

and

$$H_N = \frac{\pi^2 h^2}{\sqrt{2}w} \frac{M_1}{t_{\text{free}}} e^{-(2\sqrt{2}\pi t)/w}, \tag{2.111}$$

where t_{free} is the thickness of the ferromagnetic layer.

2.5   Micromagnetic simulation

Magnetism plays an important role in the technology of memory and information storage. However, the magnetization process is complicated. One
approach we can use to understand the phenomenon is to solve the Landau–
Lifshitz dynamic equations of motion [14, 31, 32]. The dynamic equation of
motion is a macroscopic extension of the spin Hamiltonian that can be written as
follows:

$$\frac{\partial M(r)}{\partial t} = -\gamma M(r) \times H_{eff}(r) - \frac{\lambda}{M} M(r) \times [M(r) \times H_{eff}(r)],   \tag{2.112}$$

where the effective field is the negative derivative of the total energy with respect
to the magnetization:

$$H_{eff}(r) = -\frac{\partial E_{tot}(r)}{\partial M(r)}.   \tag{2.113}$$

The first term in Eq. (2.113) represents the gyromagnetic spin electron motion,
and the second term is the phenomenological damping term. Figure 2.25 shows
the motion of the magnetization as it is described by the Landau–Lifshitz
dynamic equations. The first term describes the precession of magnetization
around the effective field, which provides the magnetic torque for the rotation
of magnetization. The second term describes the energy dissipation as a function
of the direction toward the effective field. After reaching the equilibrium state, the
magnetization aligns itself along the direction of the effective field.

2.5.1   Anisotropy energy

The total energy is calculated by summing the anisotropy energy (E_{an}), the exchange
energy (E_{ex}), the magnetostatic energy (E_{mag}) and the Zeeman energy (E_{ze}):

$$E_{tot}(r) = E_{an}(r) + E_{ex}(r) + E_{mag}(r) + E_{ze}(r).   \tag{2.114}$$

The anisotropy energy is written as follows:

$$E_{an} = K_1 \sin^2 \theta + K_2 \sin^4 \theta,   \tag{2.115}$$

where θ is the angle between the crystalline easy direction and the direction of the
magnetization, and K_1 and K_2 are the crystalline anisotropy constants. In most
cases, K_2 is negligible in comparison with K_1. Thus, Eq. (2.115) can be written in
another form as follows:

$$E_{an} = K_1 |\hat{k} \times \hat{m}|^2 = K_1 [1 - (\hat{k} \cdot \hat{m})^2],   \tag{2.116}$$

where $\hat{k}$ is the unit vector of the crystalline easy direction and $\hat{m}$ is the magnetization unit vector.

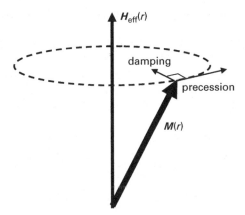

Figure 2.25. The precession of the magnetization around the effective field according to the Landau–Lifshitz dynamic equation.

2.5.2   Exchange energy

The exchange energy describes the interaction between the adjacent crystallites in the films. The exchange energy is written as

$$E_{\text{ex}} = -\frac{2A}{M^2 x^2} M \cdot \sum_j M_j, \tag{2.117}$$

where A is the effective exchange constant between the nearest-neighbor crystallites and x is the distance between the centers of two neighboring cubic cells.

2.5.3   Magnetostatic energy

The magnetostatic energy can be written as [14]

$$E_{\text{mag}} = -\frac{1}{2V} \int_V M \cdot H_{\text{mag}}(r') d^3 r', \tag{2.118}$$

where $H_{\text{mag}}(r')$ is the magnetostatic interaction field which is from the magnetostatic poles on the surface. Therefore, $H_{\text{mag}}(r')$ is

$$H_{\text{mag}}(r') = -\sum_j \int_{S_j} \hat{n}' \cdot M(r_j) \frac{r - r'}{|r - r'|^3} d^2 r^2, \tag{2.119}$$

where $\hat{n}'$ is the surface normal vector. The total magnetostatic energy is

$$E_{\text{mag}}(r_i) = -M_i \cdot \left(\sum_{j \neq i} \overset{\leftrightarrow}{D}_{ij} \cdot M_j + \frac{1}{2} \overset{\leftrightarrow}{D}_{ii} \cdot M_i \right), \tag{2.120}$$

where $\overleftrightarrow{D}_{ij}$ is referred to as the magnetostatic interaction matrix with value

$$D_{ij} = \frac{1}{V} \int\limits_{V_i} \int\limits_{S_j} \frac{(r' - r'')n'(r'')}{|r' - r''|^3} d^2 r'' d^3 r', \tag{2.121}$$

where V is the volume of the material.

2.5.4   Zeeman energy

The Zeeman energy is due to the external magnetic field acting on the magnetization; it can be written as

$$E_{ze} = -M \cdot H_{af}, \tag{2.122}$$

where H_{af} is the external applied field.

Homework

Q2.1 Equation (2.37) represents the change in magnetic moment caused by the field. Try to derive Eq. (2.37) from Eq. (2.36).

A2.1 The change of magnetic moment from Eq. (2.36) is given by

$$\Delta m = -\frac{e^2}{4m_e c} H \int \frac{R^2}{A} \sin^2 \theta \, dA,$$

where

$$dA = 2\pi R^2 \sin \theta \, d\theta.$$

The average value of R^2 is

$$\langle R^2 \rangle = r^4 \int_0^{\pi/2} \sin^3 \theta \, d\theta / 2\pi r^2 = \frac{2}{3} r^2.$$

Inserting this into Eq. (2.36), the change in magnetic moment is given by

$$\Delta m = -\frac{e^2 r^2}{6m_e c^2} H.$$

Q2.2 The Landau–Lifshitz dynamic equations of motion describe the magnetization processes under an applied field and appear in Eq. (2.112). The Landau–Lifshitz equation can be written in another form introduced by Gilbert. It is called the Landau–Lifshitz–Gilbert equation and is as follows:

$$\frac{\partial M(r)}{\partial t} = -\gamma_g M(r) \times H_{eff}(r) - \frac{\lambda_g}{M} M(r) \times \frac{\partial M(r)}{\partial t}.$$

Prove that the Landau–Lifshitz–Gilbert equation is identical to Eq. (2.112) from a mathematical point of view.

A2.2 By vector multiplying both sides of the Landau–Lifshitz–Gilbert equation by the magnetization, M, we obtain

$$M(r) \times \frac{\partial M(r)}{\partial t} = -\gamma_g M(r) \times (M(r) \times H_{\mathrm{eff}}(r)) - \frac{\lambda_g}{M} M(r)$$
$$\times \left(M(r) \times \frac{\partial M(r)}{\partial t} \right);$$

$$M(r) \times \frac{\partial M(r)}{\partial t} = -\gamma_g M(r) \times (M(r) \times H_{\mathrm{eff}}(r)) + \lambda_g M \frac{\partial M(r)}{\partial t}$$
$$- \frac{\lambda_g}{M} M(r) \left(M(r) \cdot \frac{\partial M(r)}{\partial t} \right)$$

On observing that

$$M(r) \cdot \frac{\partial M(r)}{\partial t} = 0,$$

we obtain

$$M(r) \times \frac{\partial M(r)}{\partial t} = -\gamma_g M(r) \times (M(r) \times H_{\mathrm{eff}}(r)) + \lambda_g M \frac{\partial M(r)}{\partial t}.$$

By substituting the latter equation into the right-hand side of the Landau–Lifshitz–Gilbert equation, we obtain

$$\frac{\partial M(r)}{\partial t} = -\gamma_g M(r) \times H_{\mathrm{eff}}(r) + \frac{\gamma_g \lambda_g}{M} M(r) \times (M(r) \times H(r)) - \lambda^2 \frac{\partial M(r)}{\partial t}.$$

The latter equation can be written as follows:

$$\frac{\partial M(r)}{\partial t} = -\frac{\gamma_g}{1 + \lambda_g^2} M(r) \times H_{\mathrm{eff}}(r) + \frac{\gamma_g \lambda_g}{(1 + \lambda_g^2)M} M(r) \times \frac{\partial M(r)}{\partial t},$$

which is mathematically the same as the Landau–Lifshitz equation with

$$\gamma = \frac{\gamma_g}{1 + \lambda_g^2},$$

and

$$\lambda = -\frac{\gamma_g \lambda_g}{1 + \lambda_g^2}.$$

References

[1] N. Bohr, *Phil. Mag.* **26**, 1 (1913).
[2] A. Sommerfeld, *Annalen de. Phys.* **51**(1), 125 (1916).
[3] P. Langevin, *Annales de Chim. et Phys.* **5**, 70 (1905).
[4] P. Curie, *Ann. Chim. Phys.* **5**, 289 (1895).
[5] P. Weiss, *Compt. Rend.* **143**, 1136 (1906).
[6] P. Weiss, *J. de Phys.* **6**, 661 (1907).
[7] D. Jiles, *Introduction to Magnetism and Magnetic Materials* (London: Chapman & Hall, 1989).

[8] L. Néel, *Annales de Phys.* **18**, 5 (1932).

[9] W. H. Meiklejohn and C. P. Bean, *Phys. Rev.* **102**, 1413 (1956).

[10] W. H. Meiklejohn, and C. P. Bean, *Phys. Rev.* **105**, 904 (1957).

[11] J. Nogués, and I. K. Schuller, *J. Magn. & Magn. Mater.* **192**, 203 (1999).

[12] G. W. Anderson, Y. Huai and M. Pakala, *J. Appl. Phys.* **87**, 5726 (2000).

[13] J. Nogués, D. Lederman, T. J. Moran and I. K. Schuller, *Phys. Rev. Lett.* **76**, 4624 (1996).

[14] H. N. Bertram and J. G. Zhu, *Fundamental Magnetization Processes in Thin-Film Recording Media*, Solid State Physics, Vol. 46 (New York: Academic Press, 1992), p. 299.

[15] B. D. Cullity, *Introduction to Magnetic Materials* (Reading, MA: Addison-Wesley, 1972), p. 240.

[16] W. P. Meiklejohn, *J. Appl. Phys.* **33**, 1328 (1962).

[17] D. Mauri, H. C. Siegmann, P. S. Bagus and E. Kay, *J. Appl. Phys.* **62**, 3047 (1987).

[18] A. P. Malozemoff, *Phys. Rev. B* **35**, 3679 (1987).

[19] P. Grünberg, R. Schreiber, Y. Pang, M. B. Brodsky and H. Sowers, *Phys. Rev. Lett.* **57**, 2442 (1986).

[20] S. S. P. Parkin, N. More and K. P. Roche, *Phys. Rev. Lett.* **64**, 2304 (1990).

[21] S. S. P. Parkin, R. Bhadra and K. P. Roche, *Phys. Rev. Lett.* **66**, 2152 (1991).

[22] S. S. P. Parkin, *Phys. Rev. Lett.* **67**, 3598 (1991).

[23] R. Coehoorn, *Phys. Rev. B* **44**, 9331 (1991).

[24] M. T. Johnson, S. T. Purcell, N. W. E. McGee, R. Coehoorn, J. aan de Stegge and W. Hoving, *Phys. Rev. Lett.* **68**, 2688 (1992).

[25] J. Unguris, R. J. Celotta and D. T. Pierce, *Phys. Rev. Lett.* **79**, 2734 (1997).

[26] P. A. Grünberg, *Sensors & Actuators A* **91**, 153 (2001).

[27] M. A. Ruderman and C. Kittel, *Phys. Rev.* **96**, 99 (1954).

[28] C. Kittel, *Quantum Theory of Solids* (New York: Wiley, 1963), p. 360.

[29] L. Néel and C. R. Hebd, *Seances Acad. Sci.* **255**, 1545 (1962).

[30] L. Néel and C. R. Hebd, *Seances Acad. Sci.* **255**, 1676 (1962).

[31] L. D. Landau and E. M. Lifshitz, *Sov. Phys.* **8**, 153 (1935).

[32] T. L. Gilbert, *Phys. Rev.* **100**, 1243 (1995).

3 Properties of patterned ferromagnetic films

3.1 Introduction

The first part of this chapter describes the edge effects of patterned ferromagnetic films. The edge pole of the ferromagnetic film plays a significant role in the film properties, in particular, the magnetic energy state. Since the magnetic memory cells are made of tiny pieces of patterned film or film stack, the effects due to the end poles become a dominant factor governing the stability and switching behavior of the memory cell. The second part of this chapter deals with the switching properties of a small patterned film under an external magnetic field. A coherent switching model is introduced to describe the switching properties of the film. This is the basis of the write operation of the field-MRAM cells.

3.2 Edge poles and demagnetizing field

When a ferromagnetic thin film is patterned and etched into shapes, the magnetic poles at the edge of the film are exposed. Like the end of a bar magnet, magnetic flux emits from the poles at the edge of the thin film. Inside the film, the flux points in a direction opposite to the magnetization. The magnetic field associated with the end poles is called the demagnetizing field, H_D. The magnitude of H_D is position-dependent.

Consider a semi-infinite film of thickness t and saturation magnetization M_S. The film extends from $y = 0$ in the $+y$-direction toward infinity and extends in both the $+x$- and $-x$-direction toward infinity (see Fig. 3.1). The demagnetizing field at point $(0, y_0)$ is the integral of the demagnetizing field contributed by the pole elements $dp(x)$ along the edge at $y = 0$. The end pole in a small area is given by $p_o(x)dA$. Since $dA = t\,dx$,

$$dp(x) = p_o(x)dA = 4\pi M_S t\,dx. \tag{3.1}$$

The demagnetizing field vector has an amplitude $dp(x)/r^2$ at distance $r = (x^2 + y_0^2)^{1/2}$, and a direction, n. The total H_D is given by

$$H_D = \int_{-\infty}^{\infty} \frac{dp(x)}{r^2} n. \tag{3.2}$$

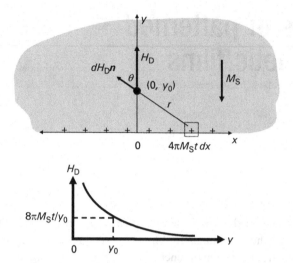

Figure 3.1. Edge poles on a semi-infinite ferromagnetic film of thickness t and demagnetizing field $\mathbf{H_D}$.

By symmetry, the demagnetizing field is inside the film plane and normal to the edge surface, or in the $+y$-direction. Vector $\mathbf{n}$ is at an angle from the edge $\sin\theta = y_0/r$, so

$$\mathbf{H_D} = (4\pi M_S t y_0) \int_{-\infty}^{\infty} \frac{dx}{r^3}. \tag{3.3}$$

The integration term equals $2/y_0^2$, since

$$\int \frac{dx}{(y_0^2 + x^2)^{3/2}} = \frac{x}{y_0^2 (y_0^2 + x^2)^{1/2}}.$$

Thus,

$$\mathbf{H_D} = \frac{8\pi M_S t}{y_0}. \tag{3.4}$$

The magnitude of the demagnetizing field $\mathbf{H_D}$ is (a) proportional to the thickness of the film and (b) inversely proportional to the distance from the edge of the film. It is in the direction opposite to $\mathbf{M_S}$. Thus, it is position-dependent, and is very strong at the edge of the film.

The demagnetizing field is not only inside the ferromagnetic film, but is also outside the film. The magnetic flux from the end pole emits in all directions. Some of that is in the air.

3.2.1   Demagnetizing factor of elliptic-shaped film

The demagnetizing field of a thin film of complex geometry cannot be calculated analytically. One needs instead to compute it using numerical techniques or to rely on measurement. For a given geometry, the relation between M_S and H_D is usually expressed as

$$H_D = NM_S(\text{in SI units}), \tag{3.5}$$

where N is called the demagnetizing factor and is dimensionless. For cgs units, M_S in Eq. (3.5) is replaced by $4\pi M_S$.

It can be proven that

$$N = N_x + N_y + N_z, \tag{3.6}$$

in an (x, y, z)-coordinate system.

In the case of an elliptic-shaped film, the demagnetizing field is uniform for a uniformly distributed magnetization [1, 2]. The demagnetization factor is a function of both the direction of magnetization and the shape of the film.

In a thin film, in which the thickness t is much smaller than the width b and length a, the demagnetizing field along the three principal axes can be approximated as follows:

$$N_a = \frac{\pi t}{4 a}\left[1 - \frac{1}{4}\frac{a-b}{a} - \frac{3}{16}\left(\frac{a-b}{a}\right)^2\right], \tag{3.7}$$

$$N_b = \frac{\pi t}{4 a}\left[1 + \frac{5}{4}\frac{a-b}{a} + \frac{21}{16}\left(\frac{a-b}{a}\right)^2\right], \tag{3.8}$$

$$N_c = 1 - N_a - N_b. \tag{3.9}$$

Figure 3.2 shows the demagnetizing factors of an elliptic-shaped thin film as a function of the length/width ratio. For a circular film, where $a = b$, one obtains

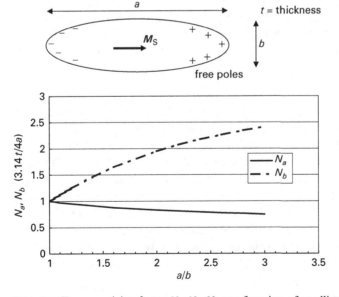

Figure 3.2. Demagnetizing factor N_a, N_b, N_c as a function of an elliptic-shaped film with length a, width b and thickness c.

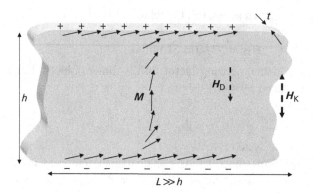

Figure 3.3. Edge curling in a patterned ferromagnetic film.

$N_a = N_b = 1$. The shape does not introduce a preferred direction to the magnetization of a circular film.

When the length to width ratio is large, however, the demagnetizing field along the width direction is much larger than along the length direction. In this case there is preferred orientation to the magnetization. Thus, the shape of the thin film, in addition to the crystalline anisotropy, affects the orientation of magnetization in a patterned thin film. In other words, the shape of the thin film provides "shape anisotropy" to the magnetization. Shape anisotropy plays an important role in the design of magnetic devices.

When a ferromagnetic film is patterned into an array of small cells in close proximity, the demagnetizing field from each cell overlaps. Thus, the net demagnetizing field becomes pattern-sensitive. When the magnetizations of all the cells are in the same direction, part of the demagnetizing field from the air is canceled and the field becomes smaller. When the magnetizations in neighboring cells are in opposite directions, the demagnetizing field is enhanced.

3.2.2   Edge curling

The demagnetizing field acts on the magnetization at the edge of the film and forces the magnetization vector to rotate toward the tangential direction of the film edge. This is usually called "edge curling" of magnetization, as illustrated in Fig. 3.3. The direction of the magnetization vector in the film is therefore position-dependent. The changes are gradual, over a distance. The magnetization in the center of the film experiences a low demagnetizing field, the direction of which is dictated by the crystalline anisotropy. The exact angle is dictated by (a) the lowest energy state of a combination of exchange energy, which tries to align the neighboring magnetization in the same direction, (b) the crystalline anisotropy energy and (c) the demagnetizing field. The first two are material-dependent, whereas the latter is shape-dependent.

Local magnetization in a patterned film changes orientation such that the magnetostatic energy is lowered. Edge curling can be viewed as a process for

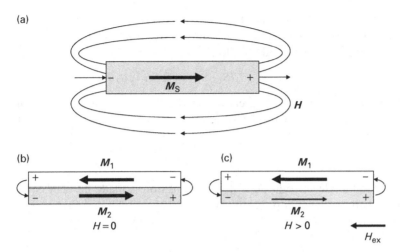

Figure 3.4. (a) Magnetic flux emitted from end poles of a single-domain bar-shaped film into the air. (b) Two-domain film under zero external field. (c) Two-domain film under non-zero external field. Note that the domain wall shifts.

the film to achieve lower total energy states. As will be shown in Section 3.3, the formation of domain is another process through which energy states can be lowered.

3.3   Magnetic domain

It has long been observed that ferromagnetic materials frequently break up into many regions, each with a different orientation of magnetization, M_S. Such a region of single orientation of magnetization is called a "domain." Figure 3.4 illustrates the domains in a ferromagnetic material and the magnetic flux outside the magnetic material. The two-domain configuration reduces the spatial extension of the demagnetizing field.

Domains are separated by a transition region called the domain wall. When an external field H is applied to the material, domains with M_S in the same direction as the H grow in size, at the expense of those with M_S in the opposite direction. The position of the wall appears to change. This process lowers the total energy of the material.

Comparing with a full film (infinite-size film without edges), the patterned film is in a higher-energy state due to the existence of edge poles. Thus, a patterned ferromagnetic film is more likely to form domains.

3.3.1   Transition region between domains: domain wall

The transition region between two regions of different magnetization orientations is called the domain wall. The wall between two antiparallel magnetization regions is called a 180-degree wall. Within the wall, the spin of the atomic plane changes the orientation gradually layer by layer. As a result, the exchange energy in the wall arises, which tries to hold the spin inside the domain wall in the same direction. This behavior can be described as follows:

$$\varepsilon_{ex} = -2JS_1 \cdot S_2 = -2J|S^2|\cos\theta \qquad (3.10)$$

where J is the exchange integral and S_1, S_2 are the spin vectors of a pair of neighboring atomic planes. The pair make a small angle θ. For small θ, $\cos\theta \sim 1 - 0.5\theta^2$, or

$$\varepsilon_{ex} \approx J|S^2|\theta^2. \qquad (3.11)$$

The total angle of rotation is π. The transition is made up of N equal steps, the angle between neighboring pairs of spin is π/N, and the exchange energy per pair of atoms is given by

$$\varepsilon_{ex} = JS^2(\pi/N)^2. \qquad (3.12)$$

The total exchange energy of a line of $N + 1$ atoms is given by

$$N\varepsilon_{ex} = JS^2\pi^2/N. \qquad (3.13)$$

The wall would thicken without limit were it not for the anisotropy energy, which acts to limit the width of the transition layer. The spins contained within the wall are largely directed away from the easy axes of the material, so there is an anisotropy energy associated with the wall, roughly proportional to the thickness.

Consider a wall parallel to the cube face of a simple cubic lattice with separating domains magnetized in opposite directions. Our goal here is to determine the number N of atomic planes contained within a wall. The energy per unit area of the wall is the sum of contributions from the exchange and anisotropy energies: $\varepsilon_w = \varepsilon_{ex} + \varepsilon_{anis}$. Let the lattice constant be denoted by a. There are $1/a^2$ lines per unit area. Thus, $\varepsilon_{ex} = (\pi^2JS^2/Na^2)$. The anisotropy energy is of the order of the anisotropy constant times the thickness, Na, or

$$\varepsilon_{anis} = KNa, \qquad (3.14)$$

where K is the crystalline anisotropy. Therefore the wall energy is the sum of the two, i.e.

$$\varepsilon_w = \frac{\pi^2JS^2}{Na^2} + KNa. \qquad (3.15)$$

There is a minimum with respect to N when

$$\frac{\partial\varepsilon_w}{\partial N} = 0 = -\left(\frac{\pi^2JS^2}{N^2a^2}\right) + Ka \qquad (3.16)$$

or

$$N = \left(\frac{\pi^2JS^2}{Ka^3}\right)^{1/2}. \qquad (3.17)$$

(a)

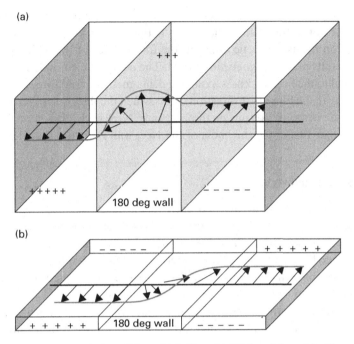

(b)

Figure 3.5. (a) Bloch wall in a thick film; (b) Néel wall in a thin film.

For iron, in which $N \sim 300$ and $a = 0.287$ nm, the wall width is ~ 86 nm. The total wall energy per unit area on this model is $\varepsilon_w = 2\pi(KJS^2/a)^{1/2}$; in iron, $\varepsilon_w \sim 1$ erg/cm^2 [3].

As illustrated in Fig. 3.4(b) and (c), the net magnetization in a multiple domain film changes gradually with an east-axis field by slowly shifting the position of the domain wall. Patterned ferromagnetic films of thickness less than 10 nm and dimension less than 100 nm cannot sustain a domain wall, since the wall is wider than the film. Without a domain wall, their net magnetizations react to an easy-axis field very differently; instead of changing gradually, they switch abruptly at a threshold field. This single-domain switching behavior is the basis of MRAM technology, in which the data storage element is a ferromagnetic film stripe. Data are stored in the MRAM cell as the orientation of the magnetization in the stripe, which is switched by an external field or other means. As the size of the ferromagnetic film is scaled down in order to increase the density of memory, this single-domain switching property continues to hold.

3.3.2   Bloch wall and Néel wall

There are two ways in which the magnetization rotates 180 degrees in the domain wall. A wall is called a Bloch wall when the magnetization rotates out of the film plane; it is called a Néel wall when the magnetization rotates in the film plane (see Fig. 3.5). Bloch walls are observed in thicker films, typically in the order of microns, but not in thin films. The reason is that the out-of-plane component of magnetization forms magnetic poles on the film surface of the region of domain wall, and results in a demagnetization field. The demagnetization field is inversely proportional to the film thickness. A strong demagnetizing field forces the magnetization to stay in the film plane direction. Thus, when the thickness of ferromagnetic films is in the order of 100 nm or thinner, only Néel walls exist. For film thicknesses less than 10 nm, the thickness of the Néel wall is greater than 1 μm [4].

3.3.3   C-state, S-state and vortex

The demagnetizing field at the two ends of an elliptic-shaped film forces the magnetization away from the long axis. As a result of symmetry, the magnetization at the end either tilts to the left or to the right. The energy levels of these two states are equal [5]. In one state, the magnetization on the two ends tilts toward the opposite direction, which is called the S-state, since the magnetization from one end of the film to the other resembles a letter "S." The other state, having the magnetization in the same direction on the two ends, is called the C-state, for the same reason. Figure 3.6 illustrates the magnetization distribution of the S-state and the C-state. Thus, low-angle domain walls form in patterned ferromagnetic film.

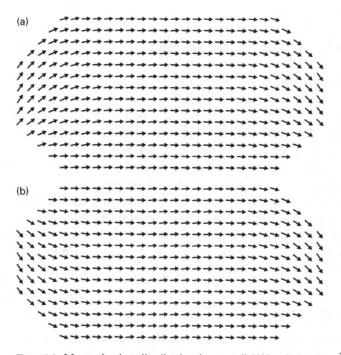

Figure 3.6. Magnetization distribution in a small ($420 \times 210 \times 3\,nm^3$) oval-shaped ferromagnetic film: (a) C-state; (b) S-state. (After ref. [5].)

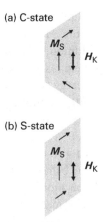

(a) C-state

M_S H_K

(b) S-state

M_S H_K

Figure 3.7. Magnetization states in an asymmetric-shaped ferromagnetic film: (a) C-state; (b) S-state.

vortex

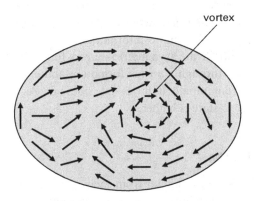

Figure 3.8. Vortex in a patterned ferromagnetic film.

On the other hand, in an asymmetric-shaped film, the lack of symmetry along the longitudinal axis changes the energy level of the S-state and the C-state. There is a preferred remnant state, which is the lower-energy state of the two. Figure 3.7 shows the state and the shape. Having only one remnant state, the variation of switching threshold is reduced.

One interesting magnetization distribution in patterned films is a vortex, as shown in Fig. 3.8. A vortex state is a state that magnetic memory device designers would like to eliminate.

3.4   Magnetization behavior under an external field

In the following few sections, we will study the behavior of magnetization under an applied external field. The angle of the magnetization in a film may rotate or switch. The mathematical treatment is greatly simplified when we view the magnetization in the entire film as a single-direction vector, M_S. In other words, edge curling and domains are not considered. In addition, the mathematics is greatly simplified by treating the demagnetizing field as a constant, H_D, independent of position. Thus, the dependence of film energy on the angle of M_S of the entire film can easily be calculated. The angle of M_S is obtained by finding the lowest energy point. This model is called the *uniform rotation model* or the *coherent rotation model*.

The uniform rotation model, although not accurately representing a patterned film, provides a first-order description of magnetization behavior in patterned films under external fields. Hence, we will proceed with this model in this chapter.

For a more detailed analysis, one needs to treat the film as many small elements, where M_S in each element does not rotate or switch coherently. This analysis is called the *micromagnet* model. Such a model is used for more complex behavior of patterned films and film stacks. For example, the switching behavior of the C-state (or the S-state) of an asymmetric-shaped film in general cannot be predicted with the uniform rotation model. Micromagnetic simulation must be carried out to analyze the switching behavior. For more information see ref. [6]. When an easy-axis field is applied to switch an asymmetric-shaped film initially in the C-state, the magnetization goes through a vortex formation and an annihilation process before switching. The film initially in the S-state does not. As a result, the switching field of a film in the C-state is higher.

3.4.1   Magnetization rotation in a full film

For a very large size film, the film edges are far from each other. In the absence of an external magnetic field, the crystalline anisotropy of the film determines the direction of magnetization M_S. The energy state of the film is lowest when M_S lies along the direction of crystalline anisotropy, H_K, which is the easy axis. The direction orthogonal to the easy axis is called the hard axis. The easy axis and the hard axis are illustrated in Fig. 3.9.

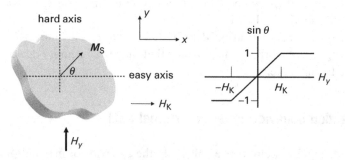

Figure 3.9. Magnetization rotates under a hard-axis field.

When an external field, H_y, is applied in the hard-axis direction, M_S responds to the external field by rotation to lower the energy state. This is illustrated in Fig. 3.9. In this case, the energies involved are the crystalline anisotropy energy, $-0.5H_K \cdot M_S \cos^2\theta$, and the magnetostatic energy, $H_y \cdot M_S \cos(\pi/2 - \theta)$, where θ is the angle between the easy axis and M_S, or

$$\begin{aligned} \varepsilon &= -0.5H_K \cdot M_S \cos^2\theta - H_y \cdot M_S \cos(\pi/2 - \theta) \\ &= -0.5H_K \cdot M_S \cos^2\theta - H_y \cdot M_S \sin\theta. \end{aligned}$$ (3.18)

Note that ε is the density of energy in a unit volume of the ferromagnetic material. The torque on M_S is $d\varepsilon/d\theta$, and, at zero torque,

$$d\varepsilon/d\theta = H_K \cdot M_S \cos\theta \cdot \sin\theta - H_y \cdot M_S \cos\theta = 0,$$ (3.19)

or

$$\sin\theta = \frac{H_y}{H_K}$$ (3.20)

The angle θ of M_S is given by

$$\theta = \sin^{-1}\left(\frac{H_y}{H_K}\right) = \sin^{-1}\left(\frac{M_y}{M_S}\right).$$ (3.21)

Note that the zero-torque point is essentially the same as the lowest-energy point. One may obtain the angle position of M_S using either approach.

3.4.2 Magnetization rotation in a patterned film

For a small patterned film, the magnetization lies along an easy direction, which is determined by the shape and the crystalline anisotropy. Consider a film stripe, as illustrated in Fig. 3.10, in which the length and the crystalline anisotropy direction

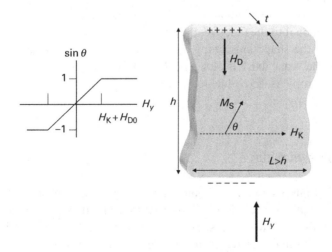

Figure 3.10. Rotation of magnetization in patterned film under an external field.

are both in the x-direction. The easy axis is in the same direction as the length of the stripe. When applying a magnetic field H_y in the y-direction, M_S rotates toward the +y-direction. As a result, edge poles appear on the edge of a film stripe with density proportional to the y-component of M_S, or $|M_S| \sin \theta$, where θ is the angle between M_S and H_K. The edge poles produce a demagnetizing field H_D, in a direction opposite to the applied field H_y and magnitude

$$H_D = H_{D0} \sin \theta. \tag{3.22}$$

In the elliptical approximation, the demagnetization field is given by

$$H_{D0} \sim (4\pi M_S t / h), \tag{3.23}$$

where t is the film thickness and h is the stripe height of the film.

In the film, the net field is given by $H_y - H_D$. Following the analysis of a full film in Section 3.4.1, one obtains

$$H_y - H_D = H_K \sin \theta. \tag{3.24}$$

Thus, the magnetization rotates through an angle of θ from the easy axis, and

$$\sin \theta = \frac{H_y}{H_K + H_{D0}}. \tag{3.25}$$

Thus,

$$\theta = \sin^{-1} \left(\frac{H_y}{H_K + H_{D0}} \right). \tag{3.26}$$

Equations (3.26) and (3.21) are identical, if we treat $(H_K + H_{D0})$ as H_K', the anisotropy of a patterned film. Thus, the equivalent anisotropy of a patterned film is usually the combination of H_K (the crystalline anisotropy) and H_D (the shape anisotropy).

Since H_D is linearly proportional to the film thickness and the material M_S and inversely proportional to the film size, the anisotropy of a small size patterned film is dominated by the shape anisotropy. For an elliptic- or oval-shaped film, the anisotropy, or the "easy axis" is along the long axis. For a circular-shaped film, the shape anisotropy is zero due to symmetry, and thus is the same as the crystalline anisotropy.

3.5   Magnetization switching

As shown in Sections 3.4.1 and 3.4.2, under a very high hard-axis field, M_S rotates to an angle $\pi/2$ from the easy axis, or aligns itself to the hard axis. When the field is removed, M_S falls back to the easy axis. By symmetry, M_S may fall in either direction. Thus, M_S may not return to its initial angle. When that happens, M_S switches its direction. Thus, a strong hard-axis field may switch the film.

3.5.1   Magnetization rotation and switching under a field in the easy-axis direction

So, M_S may be switched under an easy-axis field. But, the field strength required to switch a film is different from that of a hard-axis field. In Sections 3.4.1 and 3.4.2, we approached the analysis by applying the concept of minimum torque. In this case, we approach the analysis employing the concepts of both the minimum-energy state and minimum torque. The two approaches should yield the same results.

Consider the case illustrated in Fig. 3.11. The figure shows the energy as a function of the angle θ between M_S and the anisotropy axis, or easy axis. The lowest-energy points are at $\theta = 0$ and π. The two low-energy points are separated by a potential barrier with magnitude of $H_K M_S$, which is the maximum-energy state at $\theta = \pi/2$.

Initially, M_S is along the $+x$-direction ($\theta = 0$). An external field H_x is applied in the $-x$-direction. What is the relation between H_x and θ? How large a field is needed to switch M_S from $\theta = 0$ to π?

The energy of the film includes a new term, the magnetostatic energy term, from H as follows:

$$\varepsilon = -0.5 H_K M_S \cos^2 \theta - H_x M_S \cos(\pi - \theta)$$
$$= -0.5 H_K M_S \cos^2 \theta + H_x M_S \cos \theta. \tag{3.27}$$

Note that M_S rotates away from the easy axis and stays at an angle where the torque on M_S is zero. When the torque, $d\varepsilon/d\theta$, is zero, or

$$H_K M_S \sin \theta \cos \theta - H_x M_S \sin \theta = 0, \tag{3.28}$$

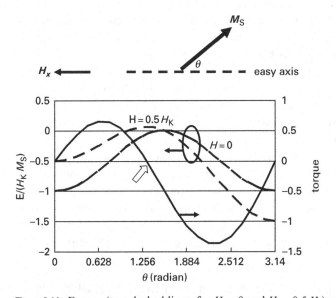

Figure 3.11. Energy (two dashed lines, for $H_x = 0$ and $H_x = 0.5\, H_k$) and torque (solid line) as a function of the angle of magnetization. The open arrow points to the zero-torque position for the case $|H_x| = 0.5\, H_K$.

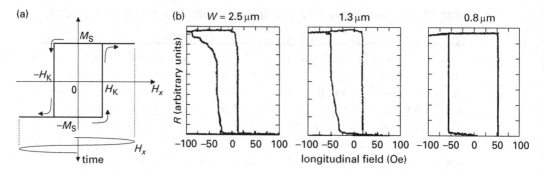

Figure 3.12. (a) *M–H* hysteresis loop. (b) Measured *R–H* loop (equivalent to *M–H* loop) of three giant magnetoresistance stripes. The vertical axis has an arbitrary scale. (After ref. [7].)

there are two angles that satisfy this equation: one is $\theta = 0$ and the other is at

$$\theta = \cos^{-1}\left(\frac{H_x}{H_K}\right). \tag{3.29}$$

The torque curve is shown in Fig. 3.11. There are two zero-torque points. The point yields a stable solution when the derivative of torque satisfies

$$d^2\varepsilon/d\theta^2 < 0, \text{ or}$$
$$H_K M_S \cos 2\theta - H_x M_S \cos \theta < 0. \tag{3.30}$$

For $H = 0.5H_K$, the stable solution is given by $\theta \sim 1.3$, which is the stable angle of M_S under such an applied field. This angle increases with increasing applied field. When the applied field is equal or exceeds H_K, θ reaches $\pi/2$, the maximum-energy barrier point. Then M_S switches over to $\theta = \pi$ as the film lowers its energy state. This process is irreversible, namely, M_S does not return to its initial angle after the applied field is removed. When an external field sweeps back and forth with magnitude greater than H_K, the trajectory of $M_x - H_x$ is a loop, as shown in Fig. 3.12(a). Such an *M–H* loop is called a hysteresis loop. After the external field is removed, a remnant state ($H_x = 0$) of magnetization is either in the $+x$- or the $-x$-direction.

Under the assumption that the magnetization of the entire piece of patterned ferromagnetic film rotates in unison, the switching takes place in a single stroke. The condition may be met when the patterned film is very narrow and long. Figure 3.12(b) shows the magnetization switching of three patterned films of different width. The narrowest (0.8 µm wide) strip exhibits switching of the single-step film, while the wider stripes exhibit multiple-step switching.

3.5.2   Magnetization rotation and switching under two orthogonal applied fields

The switching threshold is different when both $-H_x$ and H_y are applied to the film. We start from the energy equation,

$$\varepsilon = -0.5 H_K \, M_S \cos^2 \theta - H_x \, M_S \cos(\pi - \theta) - H_y \, M_S \cos(\pi/2 - \theta),$$

or

$$\varepsilon = -0.5 H_K \, M_S \cos^2 \theta + H_x \, M_S \cos \theta - H_y \, M_S \sin \theta. \tag{3.31}$$

By solving these equations for zero torque, $d\varepsilon/d\theta = 0$, the angle θ of magnetization M_S can be determined. In some cases, there is more than one solution, since both the energy minimum and the energy maximum are zero-torque points. The energy minimum is at $d^2\varepsilon/d\theta^2 < 0$ and the energy maximum is at $d^2\varepsilon/d\theta^2 > 0$. The former is stable and the latter is unstable. These two equations can be written as follows:

$$d\varepsilon/d\theta = 0.5 H_K \, M_S \sin 2\theta - H_x M_S \sin \theta - H_y M_S \cos \theta = 0, \tag{3.32}$$

$$d^2\varepsilon/d\theta^2 = H_K \, M_S \cos 2\theta - H_x M_S \cos \theta + H_y M_S \sin \theta = 0. \tag{3.33}$$

The condition that Eq. (3.33) $= 0$ is the threshold between a reversible rotation, Eq. (3.33) < 0, and irreversible switching, Eq. (3.33) > 0. For the reversible condition, once the applied field is removed, M_S returns to its initial position. Otherwise, M_S switches to a new low-energy position and remains even after the field is removed. Solving these two equations, one finds the switching threshold of M_S to be

$$H_x^{2/3} + H_y^{2/3} = (H_K)^{2/3}. \tag{3.34}$$

We leave the derivation of the switching field Eq. (3.34), as a homework exercise. This is the so-called Stoner–Wohlfarth switching "Astroid" (Fig. 3.13). (Note that this is the exact word used in the original article, not "asteroid.") Outside the Astroid, the field strength is strong enough to switch the M_S to the opposite direction. Inside the Astroid, the strength of field is not sufficient to switch the magnetization. Once the field is removed, the magnetization returns to its initial position.

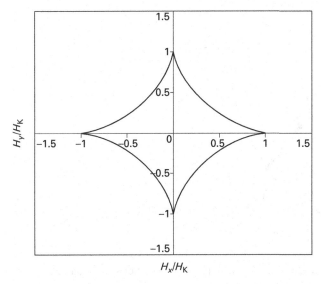

Figure 3.13. Stoner–Wohlfarth switching Astroid: $H_x^{2/3} + H_y^{2/3} = H_K^{2/3}$.

The Astroid provides a simple graphic tool for predicting reversible rotation and irreversible switching of a magnetization under a field with arbitrary direction and magnitude. Let the angle θ of M_S initially be 0. As long as the field vector H, drawn from the origin, is inside the Astroid, the magnetization M_S rotates at an angle θ less than $\pi/2$. This is independent of the angle ϕ of H. After the field is removed, M_S returns to its initial position. Figure 3.14(a) illustrates this case. If the H vector is so large that it is outside the Astroid and is in a different quadrant from M_S, M_S rotates more than $\pi/2$. As a result, after H is removed, M_S switches to the opposite direction. Figure 3.14(b) illustrates this case.

The switching Astroid also tells that, when the hard-axis field H_y is non-zero, the easy-axis switching threshold H_x is reduced, since $H_x = (H_K^{2/3} - H_y^{2/3})^{3/2}$. Figure 3.15 illustrates how the easy-axis $M\text{–}H$ loop is influenced by the hard-axis field.

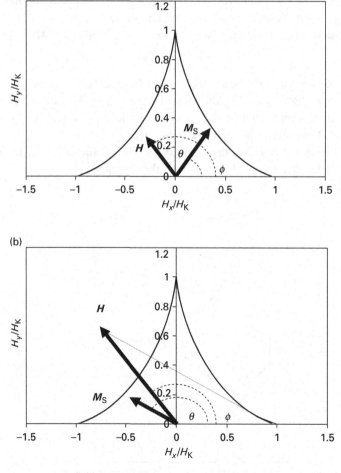

Figure 3.14. A field vector inside the Astroid (a) does not switch magnetization, whereas outside the Astroid (b) it does.

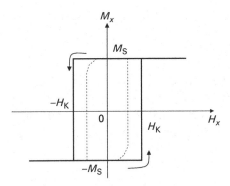

Figure 3.15. Switching of magnetization by the easy-axis field H_x, under $H_y = 0$ (solid line) and under $H_y > 0$ (dashed line).

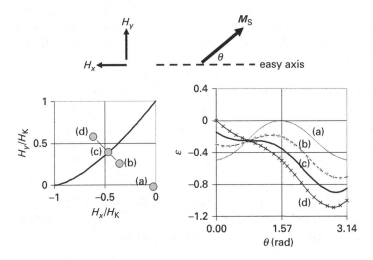

Figure 3.16. The normalized energy ε vs. the angle of magnetization for $-H_x = H_y = 0$ (a), $<H_{crit}$ (b), $= H_{crit}$ (c), $>H_{crit}$ (d), where $H_{crit} = H_K/2^{3/2}$ based on the Astroid. When $|H_x| = |H_y| = H_K$, the energy barrier $= 0$.

The Astroid switching scheme is currently being implemented to switch selectively a field-MRAM memory cell in an array with two orthogonal fields. Astroid analysis provides a good first-order estimate for samples that exhibit coherent rotation behavior at 0 K. At room temperature, thermal agitation energy also contributes to the switching of magnetization in ferromagnetic films. Thus, there is a finite probability for switching to take place even when the vector of the write field is inside the Astroid, if the field is applied for a long period of time. Similarly, there is a finite probability that switching does not happen when the field is outside the Astroid. This point can be understood from an energy diagram. The switching-energy barrier from $\theta = 0$ to π reduces as the applied field increases, and thus the switching probability increases as shown in Fig. 3.16. Even below the critical (threshold) switching field, the energy barrier is so small that the switching probability

$$P(t) \sim \exp(-E_b/k_B T) \tag{3.35}$$

is finite. For the case of zero field, point (a), the normalized switching-energy barrier is 0.5. As the external field is raised from point (a) to (b), the barrier is reduced, but still exists. At (c), the boundary of the Astroid, and (d), outside the Astroid, there is only one energy minimum where θ is near π. This point will be discussed in detail in Chapter 5.

3.6   Magnetization behavior of a synthetic antiferromagnetic film stack

Figure 3.17 illustrates a synthetic antiferromagnetic film (SAF) stack. Two ferromagnetic films are spaced by a non-magnetic film, such as a very thin ruthenium layer. The magnetizations of the two ferromagnetic layers are M_1 and M_2 and they couple through RKKY interlayer coupling. For this case, M_1 and M_2 are antiparallel.

In the absence of an external field, M_1 and M_2 both lie on the easy axis, in opposite directions. When an external field H, at an angle ϕ from the easy axis, is applied to the film stack, the energy per unit area of the film stack is the sum of the anisotropy energy, the coupling energy and the magnetostatic energy:

$$\varepsilon = -0.5H_K(M_1 t_1 \cos^2 \theta_1 + M_2 t_2 \cos \theta_2) + J_{RKKY} \cos(\theta_1 - \theta_2) \\ - H(M_1 t_1 \cos(\theta_1 - \phi) + M_2 t_2 \cos(\theta_2 - \phi)), \tag{3.36}$$

where t is the ferromagnetic film thickness, θ is the angle of the magnetization, and the subscripts 1 and 2 stand for the top and bottom ferromagnetic layer of the SAF, respectively. For the case of antiferromagnetic coupling, J_{RKKY} is positive.

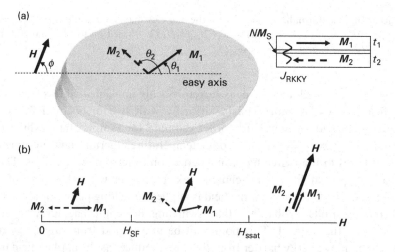

Figure 3.17. (a) Synthetic antiferromagnetic (SAF) film stack; (b) magnetization under external field.

The term H_K keeps M_1 and M_2 aligned to the easy axis, while J_{RKKY} keeps M_1 and M_2 antiparallel. The magnetostatic energy term pushes both M_1 and M_2 toward the same direction as H.

The torque on M_1 is given by

$$\Gamma_1 = \frac{\partial \varepsilon}{\partial \theta_1}, \tag{3.37}$$

and on M_2 it is

$$\Gamma_2 = \frac{\partial \varepsilon}{\partial \theta_2}. \tag{3.38}$$

At equilibrium, both torques Γ_1 and Γ_2 are zero. Solving Eq. (3.37) = 0 and Eq. (3.38) = 0, one obtains the equilibrium angular positions of θ_1 and θ_2. These equations can only be solved numerically.

For a film stack with $M_1 t_1 \sim M_2 t_2 = M_S t$, at low external field, M_1 and M_2 remain on the easy axis, just like they are in the absence of an external field. The film stack has no net moment. As illustrated in Fig. 3.17(b), when the external field is greater than a threshold, H_{SF}, both M_1 and M_2 tilt slightly toward the direction of the external field. This behavior is called *spin flop*, and the threshold field is called the spin-flop field H_{SF}. As the external field increases further, M_1 and M_2 rotate toward H, like closing a pair of scissor blades. Eventually, H reaches a value called the saturation field, H_{ssat}, under which M_1 and M_2 align themselves to H. The values of H_{SF} and H_{ssat} have been derived as follows [1]:

$$H_{SF} = \left[H_K \left(8\pi M_S N_y \frac{t}{b} - \frac{2 J_{RKKY}}{M_S t} + H_K \right) \right]^{1/2}, \tag{3.39}$$

and

$$H_{ssat} = 8\pi M_S N_x \frac{t}{b} - \frac{2 J_{RKKY}}{M_S t} - H_K, \tag{3.40}$$

where H_K is the crystalline anisotropy, a and b are length and width, respectively, and N_x and N_y are the unit-less demagnetizing factors of an elliptic-shaped film.

When the external field is greater than H_{SF}, M_1 and M_2 scissor further, their moments no longer cancel each other, and there is a net moment. As a result, the net moment of the pair aligns to the direction of the external field. When the external field rotates, the net moment of the pair follows. This is the principle of the write operation of the toggle-mode magnetic RAM. We will revisit the switching of the SAF film stack in Chapter 5.

One may also find the energy minimum using a phase-diagram technique, as illustrated in Fig. 3.18. By plotting the normalized switching energy $\varepsilon_{nor}(\theta_1, \theta_2)$ as a function of θ_1 vs. θ_2, one finds the minimum energy ε. That will be the angular direction of the M_1 and M_2 of the film stack. When $H = 0$, as shown in Fig. 3.18(a), there are two lowest-energy points at $(\theta_1, \theta_2) = (0, \pi)$ and $(\pi, 0)$. When $H > H_{SF}$, as shown in Fig. 3.18(b), there is one lowest-energy point at $(\theta_1, \theta_2) = (0.63, 1.88)$.

(a)

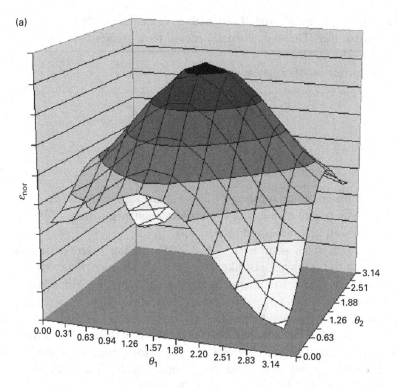

(b)

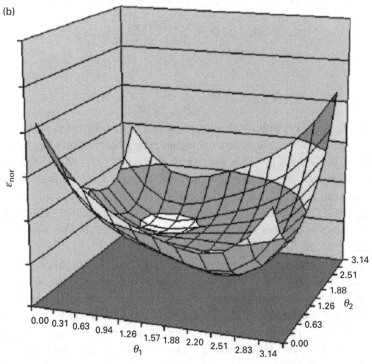

Figure 3.18. Energy of SAF film stack as a function of angles θ_1 and θ_2 of M_1 and M_2, respectively. (a) $H = 0$, (b) $H > H_{SF}$. Darker shading indicates higher energy.

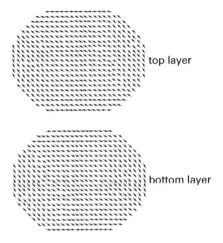

top layer

bottom layer

Figure 3.19. Simulated magnetization in a synthetic antiferromagnetic layer (after ref. [8]).

In actuality, the magnetization vector in the SAF layer is not entirely uniform. Edge curling due to the demagnetizing field still occurs, but to a lesser degree due to the partial cancelation of the demagnetizing field. Figure 3.19 shows a simulated magnetization vector distribution in an oval-shaped SAF film. Clearly, the M_S values of the top and bottom layers remain in antiparallel alignment due to the strong interlayer coupling, but the local M_S vector is not uniform.

Homework

Q3.1 Compute the demagnetizing field at the center of an infinitely long strip. The width of the strip is d. The magnetization, M_S, points (a) along the length direction and (b) along the width direction of the strip. (See Fig. 3.Q1.)

A3.1 (a) M_S is along the length direction, thus there is no pole on the film edge. The demagnetizing field $H_D = 0$.

(b) M_S is along the width direction; the demagnetizing field from one edge is given by Eq. (3.4). Summing the demagnetizing field at both edges, we obtain

$$H_D = \frac{32\pi M_S t}{d}.$$

Q3.2 Compute the demagnetizing field in the middle of a square-shaped void in an infinitely large film. The magnetization in the film is uniform and is parallel to one of the edges of the void. What is the direction of the demagnetizing field? (See Fig. 3.Q2.)

A3.2 From Eq. (3.3), integrating from $x = -d/2$ to $+d/2$, and taking $y_0 = d/2$, one obtains

$$H_D = \frac{8\sqrt{2}\pi M_S t}{d}.$$

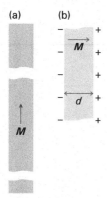

Figure 3.Q1.

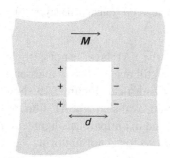

Figure 3.Q2.

Q3.3 Derive the switching Astroid of an elliptic-shaped film with length a and width b for the following two cases:

   (i) the crystalline anisotropy H_K is along the direction of the long axis of the elliptic film;

   (ii) the crystalline anisotropy H_K is along the direction of the short axis of the elliptic film.

A3.3 Hints: (i) Include the demagnetizing field in the previous analysis (i.e. the solution from Q3.2). The new $H'_K = |H_K| + H_D$, and $H_D = 4\pi M_S N_a$. The switching Astroid remains the same, except that the value of the anisotropy term is different. (ii) The same, but $|H'_K| = ||H_K| - H_D|$.

References

[1] J. A. Osborn, *Phys. Rev.* **67**, 351 (1945).

[2] E. C. Stoner, *Phil. Mag.* **36**(7), 803 (1945).

[3] C. Kittel, *Introduction to Solid State Physics*, 3rd edn (New York: John Wiley & Sons, Inc., 1968).

[4] S. Middelhoek, Ferromagnetic domain in NiFe film, Unpublished Ph.D. thesis, University of Amsterdam (1961).

[5] J.-S. Yang, C.-R. Chang, W. C. Lin and D. D. Tang, *IEEE Trans. Magnetics* **41**(2), 879 (2005).

[6] K. Ounadjala and F. B. Jenne, *Asymmetric Dot Shape for Increasing Select-Unselect Margin in MRAM Devices*, US Patent 6798691 (2003).

[7] D. D. Tang, P. K. Wang, V. S. Sperious, S. Le, R. E. Fontana and S. Rishton, *IEDM Technical Digest* (1995), p. 997.

[8] C. R. Chang and J. S. Yang, MRAM free layer magnetic state and switching behavior, Semi-annual report to TSMC, Private communication (July 29, 2004).

4 Magnetoresistance effects

4.1 Introduction

The transport property of a conduction electron in a conducting solid is predominantly governed by the electric field and the scattering events. The net effect is that the mean velocity of the electron is proportional to the electric field and is in the same direction as the electric field, which can be described as $v = \mu \cdot E$, where v is the mean electron velocity, μ is the electron mobility and E is the electric field. Here, μ is a property of the material and is independent of the direction of the electric field.

The presence of a magnetic field affects the electron transport. A moving electron experiences a Lorentz force. In SI units, the force is given by

$$F = ev \times B, \tag{4.1}$$

where B is the magnetic induction. Note that the Lorentz force F is not in the same direction as the electron velocity, but is normal to v. Thus, between each scattering, the Lorentz force "curls" the traveling path of the electrons into a helical path, which causes the electrons to travel farther. Effectively, the magnetic field increases the resistance of the material. This is called the magnetoresistance (MR) effect. It is positive, since the resistance increases. In a normal metal, the magnitude of the magnetoresistance is too small for any technological application.

In ferromagnets, however, magnetoresistance effects are much more pronounced and originate from totally different mechanisms. Listed here in their order of discovery, they are: anisotropic magnetoresistance (AMR) in the range of 2%; giant magnetoresistance (GMR) in the range up to 20%; and tunneling magnetoresistance (TMR) in the range up to 500%. These three MR effects have been the basis of the rapid growth of data storage density of magnetic hard disks. Figure 4.1 shows the latest state of the art of the MR ratio since 1995 (early research data points were not included). This chapter will describe these three MR effects.

4.2 Anisotropic magnetoresistance

In 1856, William Thomson (Lord Kelvin) was the first to discover the magnetoresistance effect. He wrote [1]

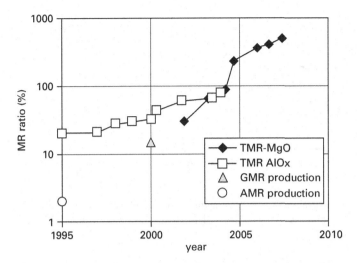

Figure 4.1. Reported magnetoresistance ratios of different types since 1995.

... had communicated to the Royal Society a description of experiments by which I found that iron, when subjected to magnetic force, acquires an increase of resistance to the conduction of electricity across, the lines of magnetization. By experiments more recently made, I have ascertained that the electric conductivity of nickel is similarly influenced by magnetism, but to a greater degree, and with a curious difference...

Kelvin discovered the dependence of resistance of iron and nickel on the amplitude and, more importantly, on the *direction* of the applied magnetic field. This effect is referred to as anisotropic magnetoresistance (AMR). The AMR effect in ferromagnetic materials, although the resistance ratio is only of the order of a few percent, is orders of magnitude greater than the MR in normal metal. The hard-disk industry successfully developed AMR sensors to replace inductive sensors as the read head in the late 1970s.

In a ferromagnet, the orientation of the orbital angular momentum is coupled to the lattice of the ferromagnet and is fixed in direction (the easy axis); its time average value is non-zero, i.e. it is unquenched There is a substantial spin-orbit coupling, which affects the scattering rate of the conduction electrons and, thus, the electrical resistance of the ferromagnet. When the magnetization is perpendicular to the current direction, the scattering cross-section is smaller, whereas when the magnetization direction is parallel to the current direction, the scattering cross-section is larger.

Referring to Fig. 4.2, the electric field E in the ferromagnetic film can be decomposed into two components: one that is parallel to the magnetization, $E_{//}$, and one perpendicular to the magnetization, $E_\perp$. Similarly, the current density J can be decomposed into $J_{//}$ and $J_\perp$. According to Ohm's Law, the relation between E and J is given by

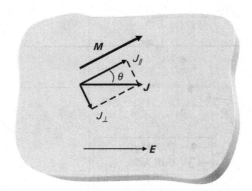

Figure 4.2. The resistance of the ferromagnet depends on the angle between the direction of current flow and the magnetization. This is called the anisotropic magnetoresistance (AMR) effect.

$$E_{//} = \rho_{//}J_{//}, \tag{4.2}$$

$$E_{\perp} = \rho_{\perp}J_{\perp}, \tag{4.3}$$

where

$$E_{//} = |E|\cos\theta, \tag{4.4}$$

$$E_{\perp} = |E|\sin\theta, \tag{4.5}$$

and

$$J_{//} = |J|\cos\theta, \tag{4.6}$$

$$J_{\perp} = |J|\sin\theta, \tag{4.7}$$

where θ is the angle between J and M. Since

$$\begin{aligned}
|E| &= E_{//}\cos\theta + E_{\perp}\sin\theta \\
&= |J|(\rho_{//}\cos^2\theta + \rho_{\perp}\sin^2\theta) \\
&= |J|(\rho_{//} - \rho_{\perp})\cos^2\theta + \rho_{\perp}.
\end{aligned} \tag{4.8}$$

the resistivity of the film is a function of the angle between the current and the magnetization θ:

$$\rho = \frac{|E|}{|J|} = (\rho_{//} - \rho_{\perp})\cos^2\theta + \rho_{\perp}. \tag{4.9}$$

The ratio $(\rho_{//} - \rho_{\perp})/\rho_{\perp}$ is called the magnetoresistance (MR) ratio. Thus, the resistance of the film is given by

$$R = R_0 + \Delta R, \tag{4.10}$$

where R_0 is the resistance when J and M are in the same direction, and

$$\Delta R = \Delta R_0 \cos^2\theta, \tag{4.11}$$

where θ is the angle between J and M.

4.3   Giant magnetoresistance

In 1988, two groups, led by A. Fert [2] and P. Grünberg [3], independently discovered the giant magnetoresistance effect (GMR) in magnetic multilayers (layers of ferromagnetic and non-magnetic metals stacked on each other). Grünberg's group used the Fe/Cr/Fe trilayer and Fert's group used $(Fe/Cr)_{60}$. The resistivity of Fe/Cr multilayers as a function of magnetic field at 4.2 K is shown in Fig. 4.3 [2]. In 2007, Fert and Grünberg were awarded the Nobel Prize in Physics for their discovery of GMR. This discovery was a great breakthrough in the field of thin-film magnetism and magneto-transport studies.

Antiferromagnetic and ferromagnetic coupling between the magnetization of the ferromagnetic layers can be adjusted by controlling the thickness of the non-magnetic layer, and an RKKY-like mechanism predicts the results that are described in Chapter 2. The resistance is higher when the magnetizations of the ferromagnetic films are antiparallel. When the magnetic layers are in a parallel alignment, conduction electrons of compatible spin type are able to move through the heterostructure with minimal scattering, and the overall resistance of the material is lowered. By changing the relative magnetization of the ferromagnetic layers from parallel to antiparallel, a large change in resistance can be measured.

The GMR effect can be understood from the following simplified picture. The electrical conductivity in metals can be described by considering two conductivity channels, corresponding to the up-spin and down-spin electrons. The difference in scattering between the antiparallel and parallel alignment multilayers can be

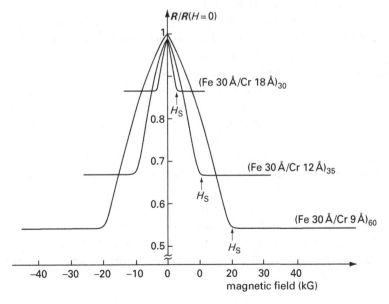

Figure 4.3. Dependence of the measured magnetoresistance on the applied magnetic field for $[Fe 30 \text{ Å}/Crx \text{ Å}]_N$ structures with $N = 30$, 35 and 60 at 4.2 K [2].

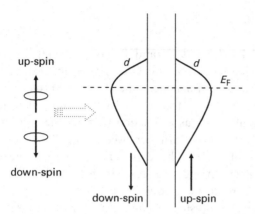

Figure 4.4. Energy bands of a non-magnetic metal at the Fermi level. The down-spin and up-spin electrons are equal in the d band.

understood using a band structure picture. As shown schematically in Fig. 4.4, in a normal metal there are equal numbers of up-spin and down-spin states at the Fermi level; therefore, up- and down-spin electrons travel through a normal metal with equal probability. In a spin-polarized metal, such as a ferromagnet, however, there are more states of one spin direction than the other at the Fermi level (Fig. 4.5). The density of states at the Fermi level is different for up-spin and down-spin. In the general case, the majority-spin electrons (in which the spin of the electrons is parallel to the direction of magnetization of the ferromagnet) have a weak scattering (see Fig. 4.5(a)). Moreover, the minority-spin electrons (in which the spin of electrons is antiparallel to the direction of magnetizations of ferromagnet) have a strong scattering. Therefore, the resistances of two spin electrons are different. Figure 4.6 shows similar magnetic multilayers to those in Fig. 4.5; however, the resistances from spin scattering represent the magnetic layers. In Fig. 4.6(a), the up-spin electrons are weakly scattered and the down-spin electrons are strongly scattered in both ferromagnetic layers. The resistance can be modeled by two small resistances $R_\uparrow$ in the up-spin channel and by two large resistances $R_\downarrow$ in the down-spin channel. Since the up-spin and down-spin resistor channels are connected in parallel, the total resistance (R_p) for the case shown in Fig. 4.6 (a) can be easily determined. On the other hand, in the antiferromagnetically aligned multilayers, shown in Fig. 4.6(b), the up-spin electrons are weakly scattered in the bottom ferromagnetic layer and are strongly scattered in the top ferromagnetic layer. The up-spin electrons are strongly scattered in the bottom ferromagnetic layer and the down-spin electrons are weakly scattered in the top ferromagnetic layer. In the antiferromagnetically aligned multilayers, the total resistance can be described as R_{ap}. Therefore, the difference in resistance between two cases is given by

$$\Delta R = R_p - R_{ap} = -\frac{1}{2}\frac{(R_\uparrow - R_\downarrow)^2}{R_\uparrow + R_\downarrow}. \tag{4.12}$$

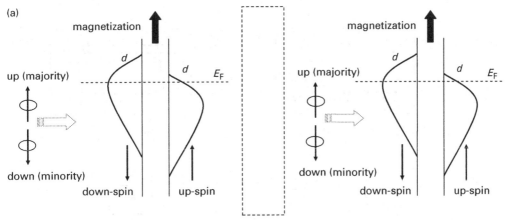

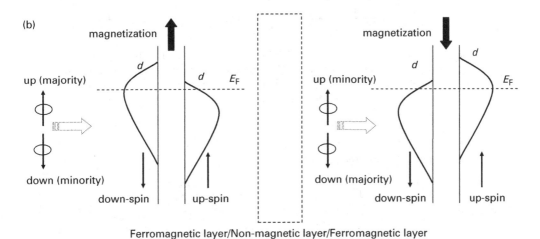

Figure 4.5. Energy bands in a FM/NM/FM structure. (a) The magnetizations of two ferromagnetic layers are parallel. (b) The magnetizations of two ferromagnetic layers are antiparallel.

From Eq. (4.12), we observe that the larger the difference between $R_\uparrow$ and $R_\downarrow$, the larger the magnetoresistance.

4.4   Tunneling magnetoresistance

Tunneling magnetoresistance (TMR) is due to spin-polarized tunneling, which was discovered by Tedrow and Meservey in 1970 [4, 5]. They used superconducting layers as detectors to measure the electron spin polarization of magnetic metals using the Zeeman-split quasi-particle density of states. In 1975, Jullière proposed the model that predicted tunneling magnetoresistance in a ferromagnetic/insulator/

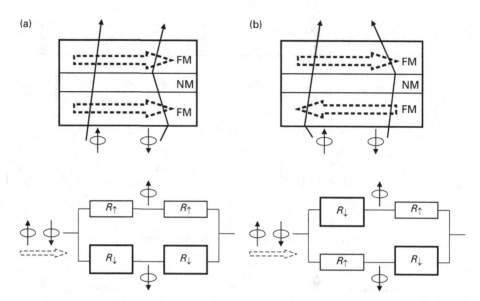

Figure 4.6. Electron transport in a FM/NM/FM structure. (a) The magnetizations of two ferromagnetic layers are parallel. The bottom picture shows the resistor network in two-channel mode. (b) The magnetizations of two ferromagnetic layers are antiparallel. The lower parts of the diagram show the resistor network in two-channel mode.

ferromagnetic layer structure (FM/I/FM) [6]. The Jullière model predicts the tunneling magnetoresistance (TMR) ratio to be

$$\text{TMR} = \frac{\Delta R}{R_\text{p}} = \frac{R_\text{ap} - R_\text{p}}{R_\text{p}} = \frac{2P_1 P_2}{1 - P_1 P_2}, \tag{4.13}$$

where R_ap and R_p are the resistances when the magnetizations of the ferromagnetic layers are antiparallel and parallel, respectively, and P_1 and P_2 are the spin polarizations of the two ferromagnetic layers. Jullière also reported the conductance between electrodes of Fe and Co separated by a Ge barrier. He measured the tunneling conductance dependence on voltage at 4.2 K [6]. At zero bias, the change in conductance is 14%, which reduces rapidly on increasing the DC bias to become approximately 2% at 6 mV.

According to Eq. (4.13), the polarization of Co is 0.34 and that of Fe is 0.44.

Miyazaki and Tezuka report that the TMR ratio is roughly proportional to the product of the spin polarizations of two ferromagnetic layers and they observed the TMR ratio to be 2.7% at room temperature in a NiFe/Al$_2$O$_3$/Co system [7, 8].

A giant TMR ratio in a FM/I/FM system was first observed in 1995 by Moodera *et al.* [9]. They developed a fabrication process for a smooth and pinhole-free Al$_2$O$_3$ deposition. It was shown that the TMR ratio is larger (10%), as shown in Figs. 4.7 and 4.8 [9, 10]. Table 4.1 shows the TMR ratios in different materials in FM/I/FM systems at different temperatures [7] before 1994. Nowadays, the TMR ratio is more than 70% based on Al$_2$O$_3$ barriers at room temperature.

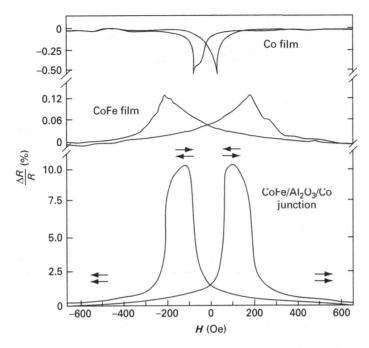

Figure 4.7. Dependence of the magnetoresistance ratio on the applied magnetic field in a Co film, a CoFe film, and a CoFe/Al$_2$O$_3$/Co structure at 295 K [9].

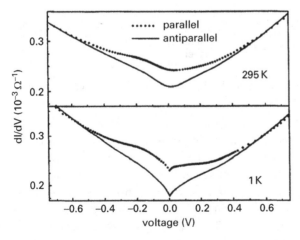

Figure 4.8. Dependence of the conductance on the bias voltage for parallel and antiparallel alignments of the magnetizations at 1 and 291 K [10].

Moreover, the TMR ratio of a FM/I/FM junction with a MgO barrier is larger than 300% at room temperature; we will discuss this breakthrough later.

The Jullière model provides an important and simple physical image with which to understand the TMR effect. In 1989, Slonczewski proposed a theory to analyze

Table 4.1 TMR ratio in FM/I/FM systems at different temperatures [7]

Year	FM/I/FM junction	TMR (%) 4.2 K	TMR (%) 300 K
1975	Fe/Ge/Co	14	
1982	Ni/NiO/Ni	0.5 (1.5 K)	
	Ni/NiO/Co	2.5	
	Ni/NiO/Fe	1.0 (2.5 K)	
1987	Ni/NiO/Co		0.96
1990	Fe–C/Al$_2$O$_3$/Fe–Ru		1
1991	Gd/GdO$_x$/Fe	5.6	
	Fe/GdO$_x$/Fe	7.7	
	Ni/Al$_2$O$_3$/Co		0.5
1993	Ni$_{80}$Fe$_{20}$/Al$_2$O$_3$/Co	5	2.7
1994	Ni$_{81}$Fe$_{19}$/MgO/Co		0.2
	Fe/Al$_2$O$_3$/Fe	30	18
	Fe$_{50}$Co$_{50}$/Al$_2$O$_3$/Co	7.2	3.5
	Fe/Al$_2$O$_3$/Co	8.5	3.3

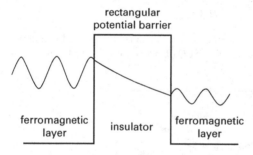

Figure 4.9. Potential barrier in a FM/I/FM structure and wave function of tunneling electron.

the transmission of spin-polarized electrons flowing through a tunneling barrier [11]. He considered the potential barrier of an insulator between two identical ferromagnetic layers to be a rectangular barrier, as shown in Fig. 4.9. The theory is based on the free-electron model, and the Schrödinger equation is solved to determine the conductance as a function of the relative magnetization alignment of the two ferromagnetic layers. The dependence of the conductance on the angle θ between the magnetizations of the two ferromagnetic layers is as follows:

$$G(\theta) = G_s(1 + P_F^2 \cos\theta), \tag{4.14}$$

where G_s is the mean surface conductance and is independent of θ, and P_F is the effective spin polarization of the tunneling electrons given by

$$P_F = \frac{k^\uparrow - k^\downarrow}{k^\uparrow + k^\downarrow} \frac{\kappa^2 - k^\uparrow k^\downarrow}{\kappa^2 + k^\uparrow k^\downarrow}, \tag{4.15}$$

where $k^\uparrow$ is the Fermi wave vector in the up-spin band, $k^\downarrow$ is the Fermi wave vector in the down-spin band and $i\kappa$ is the imaginary wave vector in the barrier. Note that κ is determined by the potential barrier V_b:

$$\kappa = \frac{1}{\hbar}\sqrt{2m(V_b - E_F)}, \tag{4.16}$$

where E_F is the Fermi energy. In the limit of a high potential barrier, κ approaches ∞. Therefore, the spin polarization from Eq. (4.15) becomes

$$\lim_{\kappa \to \infty} P_F = \frac{k^\uparrow - k^\downarrow}{k^\uparrow + k^\downarrow}. \tag{4.17}$$

Slonczewski's results reduce to Jullière's results. If the potential barrier is low, the TMR ratio decreases with decreasing potential barrier.

Slonczewski's model treats the interface of the ferromagnetic layer and the insulator more realistically and indicates that the spin polarization of the conductance is not only a property of the ferromagnets, but also relates to the quality of the insulator.

4.4.1   Giant tunneling magnetoresistance

In 2001, Butler *et al.* and Mathon and Umerski independently proposed the theoretical prediction, by first-principles calculations, of a giant tunneling magnetoresistance ratio over 1000% in fully epitaxial Fe(001)/MgO(001)/Fe(001) sandwiches [12, 13]. They found that the tunneling conductance depends strongly on the symmetry of the Bloch states and that the Bloch states of different symmetries decay at different rates in the tunneling barrier. Moreover, the decay rates are determined by the complex energy bands of symmetry in the barrier.

The interfacial structure of Fe(001)/MgO(001) is shown in Fig. 4.10. The electronic density of states is calculated for the majority- and minority-spin channels in each layer. They found that the density of states near the interface of Fe(001)/MgO(001) is quite different from that of the bulk.

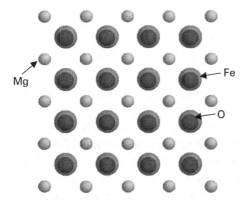

Figure 4.10. The interface of Fe(100)/MgO [12].

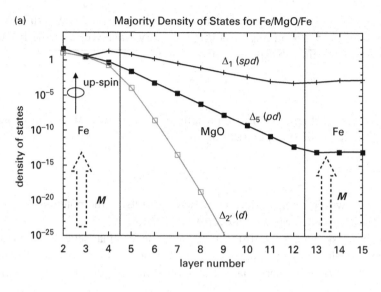

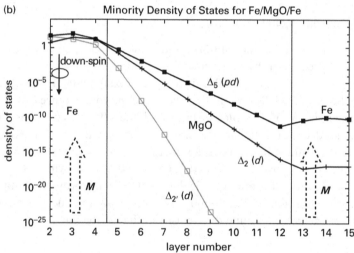

Figure 4.11. The density of states of (a) majority-spin and (b) minority-spin electrons when the magnetizations of the two Fe layers are parallel in a Fe/MgO/Fe structure [12].

The two authors defined the tunneling density of state to be under the following boundary conditions: on the left-hand side of the interface there is an incoming Bloch state with unit flux and corresponding reflected Bloch states, and on the right-hand side are the corresponding transmitted Bloch states. The density of states with Δ_1 (1 and z combination symmetry), Δ_2 ($x^2 - y^2$ symmetry), Δ_2' (xy symmetry) and Δ_5 (zx and zy combination symmetry) for parallel alignment of the magnetizations in the Fe layers with MgO are shown in Fig. 4.11 [12]. Figures 4.11 (a) and (b) show the tunneling density of state for majority- and minority-spin

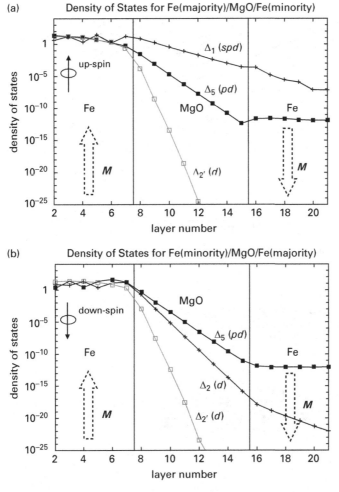

Figure 4.12. The density of states with the magnetizations of two Fe layers are antiparallel in a Fe/MgO/Fe structure [12].

electrons. Figure 4.12 shows the tunneling density of state for an antiparallel alignment of the magnetizations in the Fe layers. The density of states for the majority-spin electrons with Δ_1 symmetry decay slowly in MgO layers. However, there are no Δ_1 minority states at the Fermi level on the right-hand side of MgO; the minority-spin electrons are totally reflected. In order to obtain the tunneling conductance, the density of states are inserted into the Landauer conductance formula. Figure 4.13 shows the tunneling conductance of the majority- and minority-spin electrons, for the magnetizations of two ferromagnetic layers in parallel alignment, and the tunneling conductance for the magnetizations of two ferromagnetic layers in antiparallel alignment. It is found that the tunneling conductance of the majority-spin electrons is much larger than that of

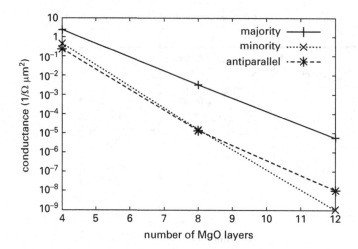

Figure 4.13. Dependence of conductance on MgO layers [12].

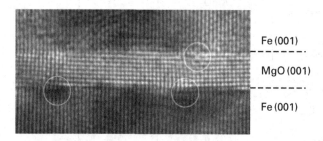

Figure 4.14. Cross-section transmission electron microscope image of Fe(001)/MgO(001) (1.8 nm)/Fe(001) MTJ [14].

the minority-spin electrons or that of the antiparallel alignment for all thicknesses of the MgO layer. According to the calculated results, the TMR ratio increases with the MgO thickness. Moreover, the TMR ratio could be over 1000% with a thicker MgO layer.

In 2004, Yuasa *et al.* and Parkin *et al.* independently proposed that the TMR ratio can be measured over 180% at room temperature [14, 15]. Yuasa used molecular beam epitaxy (MBE) to prepare single-crystal Fe(001)/MgO(001)/Fe (001) magnetic tunnelling junction (MTJ). The cross-section transmission electron microscope image of a Fe(001)/MgO(001)(1.8 nm)/Fe(001) MTJ is shown in Fig. 4.14 [14]. The image shows that the MgO(001) lattice is elongated along the [001] axis and is compressed along the [100] axis to match the Fe lattice. Figure 4.15 shows the magnetoresistance curves of Fe(001)/MgO(001)/Fe(001) junctions at 20 and 293 K. The results show that the TMR ratio is 247% at 20 K and 180% at 293 K.

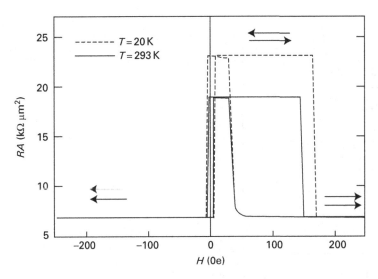

Figure 4.15. Measured resistance curves in Fe/MgO/Fe structure at 20 and 293 K [14].

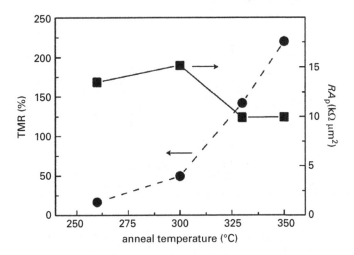

Figure 4.16. Dependence of magnetoresistance ratio and RA on anneal temperature [15].

Parkin demonstrated that the TMR ratio could exist up to 220% at room temperature and 300% at low temperature in the sputter-deposited polycrystalline MTJs grown on an amorphous underlayer, but with a highly oriented (001) MgO tunnel barrier and CoFe electrodes. He also showed that the TMR ratio is increased by thermal annealing in a vacuum at high temperature. The dependence of the TMR ratio on the anneal temperature in a 75 Å$(Co_{70}Fe_{30})_{80}B_{20}/20$ Å MgO/ 35 Å $Co_{70}Fe_{30}$ MTJ junction is shown in Fig. 4.16 [15], and the TMR ratio can be up to 220% after annealing at 350 °C.

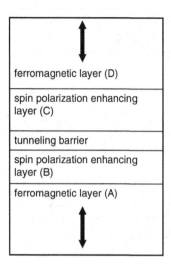

Figure 4.17. Structure of a perpendicular MTJ.

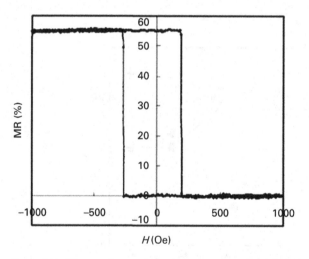

Figure 4.18. Measured magnetoresistance curve of $CdFeCo/CoFe/Al_2O_3/CoFe/TbFeCo$ perpendicular MTJ structure at room temperature [16].

4.4.2   Tunneling magnetoresistance in perpendicular magnetic tunneling junction

All MTJs we have mentioned before are longitudinal, in which the magnetization is oriented in the film plane. In recent years, MTJs with magnetization perpendicular to the film plane have attracted a lot of attention because of their application potential.

Figure 4.17 shows a schematic of a perpendicular MTJ. The ferromagnetic layers, A and D, have strong perpendicular anisotropy. Therefore, the magnetizations in the A and D ferromagnetic layers are perpendicular to the film plane.

Moreover, the interaction between ferromagnetic layer A and the spin polarization enhancing layer B is strongly ferromagnetic coupled. The interaction between ferromagnetic layer D and the spin polarization enhancing layer C is also strongly ferromagnetic coupled. The magnetizations within the spin polarization enhancing layers B and C are also perpendicular to the film plane.

Figure 4.18 shows the tunneling magnetoresistance curve in a AlCu 25 nm/GdFeCo 50 nm/CoFe 1 nm/Al_2O_3/CoFe 1 nm/TbFeCo 30 nm/Pt 2 nm MTJ structure [1].

The TMR ratio is 55% in a perpendicular MTJ at room temperature. The perpendicular MTJ also can be successfully demonstrated with an MgO barrier [17]. The TMR ratio is 64% in a GdFeCo 100 nm/Fe 3 nm/MgO 3 nm/Fe 3 nm/TbFeCo 100 nm structure at room temperature [16].

Homework

Q4.1 The ΔR between the parallel and antiparallel alignments of magnetizations in a FM/NM/FM trilayer is shown in Eq. (4.12). Solve this equation using the two-channel model.

A4.1 The resistance R_{ap} can be written as follows:

$$R_{ap} = \frac{R_\downarrow + R_\uparrow}{2},$$

and Rp can be written as:

$$R_p = \frac{1}{\dfrac{1}{2R_\downarrow} + \dfrac{1}{2R_\uparrow}} = \frac{2R_\downarrow R_\uparrow}{R_\downarrow + R_\uparrow}.$$

Therefore, ΔR is given by

$$\Delta R = R_P - R_{ap} = -\frac{1}{2} = \frac{(R_\uparrow + R_\downarrow)^2}{R_\uparrow + R_\downarrow}.$$

References

[1] W. Thomson, *Proc. Roy. Soc. London* **8**, 546 (1857).

[2] M. N. Baibich, J. M. Broto, A. Fert, *et al. Phys. Rev. Lett.* **61**, 2472 (1988).

[3] G. Binasch, P. Grunberg, F. Saurenbach and W. Zinn, *Phys Rev. B* **39**, 4828 (1989).

[4] R. Meservey, P. M. Tedrow and P. Fulde, *Phys. Rev. Lett.* **25**(18), 1270 (1970).

[5] P. M. Tedrow and R. Meservey, *Phys. Rev. Lett.* **26**(4), 192 (1971).

[6] M. Jullière, *Phys. Lett.* **54A**(3), 225 (1975).

[7] T. Miyazaki and N. Tezuka, *J. Magn. & Magn. Mater.* **151**, 403 (1995).

[8] T. Miyazaki and N. Tezuka, *J. Magn. & Magn. Mater.* **139**, L231 (1995).

[9] J. S. Moodera, L. R. Kinder, T. M. Wong and R. Meservey, *Phys. Rev. Lett.* **74**(16), 3273 (1995).

[10] J. S. Moodera, J. Nowak and R. J. M. van de Veerdonk, *Phys. Rev. Lett.* **80** 2941 (1998).

[11] J. C. Slonczewski, *Phys. Rev. B* **39**, 6995 (1898).

[12] W. H. Butler, X. -G. Zhang, T. C. Schulthess and J. M. MacLaren, *Phys. Rev. B* **63**, 054416 (2001).

[13] J. Mathon and A. Umerski, *Phys. Rev B.* **63**, 220403 (2001).

[14] S. Yuasa, T. Nagahama, A. Fukushima, Y. Suzuki and K. Ando, *Nat. Mater.* **3**, 868 (2004).

[15] S. S. P. Parkin, C. Kaiser, A. Panchula, P. M. Rice, B. Hughes, M. Samant and S. -H. Yang, *Nat. Mater.* **3**, 862 (2004).

[16] H. Ohmori, T. Hatori and S. Nakagawa, *J. Appl. Phys.* **103**, 07A911 (2008).

[17] N. Nishimura, T. Hirai, A. Koganei, T. Ikeda, K. Okano, Y. Sekiguchi and Y. Osada, *J. Appl. Phys.* **91**, 5246 (2002).

5 Field-write mode MRAMs

5.1 Introduction

Magnetic ferrite core memory was invented and produced in the 1960s, prior to semiconductor memory. Ferrite cores are made from a paste of ferrite powers, which are sintered at high temperature. The process of forming a discrete core is not as scalable as the integrated circuit process on a silicon wafer. The product life of a magnetic core was short, and in the 1970s this technique was replaced by semiconductor memory. A similar fate happened to magnetic bubble memory, another type of magnetic memory, which was built on a magnetic garnet material substrate (gadolinium gallium garnet, $Gd_3Ga_2(GaO_4)_3$). The bit density of bubble memory technology is scalable since it is made with a planar process, similar to the silicon integration circuits. However, because it is on garnet, it is passive and cannot perform logic functions (such as address decoding), and it requires a companion silicon chip to provide the logic function to complete the memory access function. Even with better memory performance, magnetic bubble memory could not compete against magnetic hard disk and semiconductor memory, which continue to show a clear path of scaling for a lower cost. By the mid 1980s, commercial magnetic bubble production had ended.

Subsequent efforts in the development of magnetic memory have been focused on the integration of magnetic thin-film memory devices into silicon wafer processes. Magnetic memory devices exist in the form of thin-film stacks, which can easily be integrated into the back-end metal wiring metallurgy process. The technological barrier of introducing magnetic material into the silicon process becomes much smaller. Thus, a magnetic memory device may take advantage of the semiconductor industry's scaling law for cost reduction. The new generation of integrated magnetic memory is more scalable and possesses properties that are not available in any of today's semiconductor memories, namely non-volatility, infinite endurance and fast data access.

In the late 1980s, many attempts were made to construct memory devices based on the anisotropic magnetoresistance (AMR) effect and the giant magnetoresistance (GMR) effect [1]. Many patents were filed (for just a few, see refs. [2], [3] and [4]). A short history outlining these attempts can be found in ref. [5].

The result of one such attempt [6] is shown in Fig. 5.1. The storage element is a GMR film stripe, with the magnetization of one ferromagnetic layer pinned to a

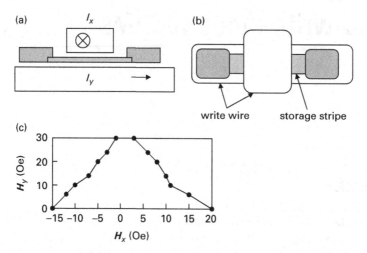

Figure 5.1. (a) Side view and (b) top view of MRAM consisting of GMR storage stripe and two orthogonal write lines. (c) The switching threshold under two orthogonal external fields. (After ref. [6], with permission from the IEEE.)

fixed direction, while that of the other is free to switch under an external field. When the width of the GMR film stripe is reduced to sub-micron dimensions, the stripe exhibits binary resistance states, without any intermediate states. A memory cell is made up of a GMR data storage element and two metal wires in orthogonal directions for writing data into the storage element. A cell in a two-dimensional cell array is selectively written based on the Astroid switching principle described in Chapter 3. The magnetic fields of two orthogonal current-carrying wires switch the free layer of the GMR stripe that is located at the intersection. The stored data are read out by sensing the resistance of the storage element. Thus, data contained in magnetic RAMs is randomly accessible. This concept remains the basic operational principle of many of today's field-MRAMs.

There are, however, serious drawbacks in AMR and GMR film memory devices. For example, they cannot provide a sufficient read-back signal due to their relatively low magnetoresistance ratio (AMR ~3%, GMR <15–20%) and, more importantly, due to their low metal film resistance, typically of the order of tens of $\mu\Omega$/sq. To generate a sufficient read signal (say, >50 mV) to be detected by a CMOS sensing circuit, the read current must be in the range of tens of milliamps. This memory concept again failed due to the impedance mismatch between CMOS and AMR/GMR devices.

Meanwhile, magnetic tunnel junction (MTJ) devices were being developed for memory and hard-disk-drive applications [7–10]. The resistance of the MTJ device is in the kilo-ohm range, much larger than the metal film resistance of AMR and GMR devices. Thus, the resistance of MTJs is of the same order as the source-drain resistance of a MOSFET. The current across the tunnel junction is proportional to the junction area, and thus approximates well the scaling of a MOSFET. In addition, the tunneling magnetoresistance ratio (TMR) is one or two orders of magnitude greater than those of GMR or AMR devices.

The technological potential of MTJs was thus recognized, and consequently research efforts in both academia and the hard-disk industry have been very intensive. While progress was continuously made to raise the TMR ratio of the AlO_x tunnel barrier, the introduction of the MgO tunnel barrier with a CoFeB ferromagnetic thin film drastically raises the TMR ratio to giant TMR values. As shown in Fig. 4.1, by 2007 the maximum giant TMR reported in the literature reached nearly 500%. That makes the read-back signal much larger and eases the read-back circuit design constraints. Understanding the potential of TMR devices, the semiconductor industry began to develop MRAM devices based on MTJs. A short history of their early development can be found in ref. [11].

Like GMR memory devices, MTJ memory devices store binary data (0, 1) as the orientation of the magnetization in the free layer, which may be parallel or antiparallel to the pinned layer. There are many ways to write data into the MTJ device, i.e. to switch the free-layer magnetization. The MRAM in which data are written into MTJ via a magnetic field is called field-MRAM. A current in a wire adjacent to the MTJ generates the write field. The two most mature field-MRAM technologies are Astroid-mode and toggle-mode. This chapter will focus on these two field-MRAMs. Other field-MRAMs will be briefly described at the end of this chapter.

In addition to the field-MRAM, spin-torque-transfer mode MRAM (STT-MRAM) is a potentially higher-density, lower-power memory. Rather than switching the free layer of the MTJ with magnetic fields from adjacent wires, a spin-polarized current through the MTJ switches its free layer. The STT-MRAM will be the subject of Chapter 6.

5.2   Magnetic tunnel junction RAM cell

5.2.1   Cross-point array

In the early days of the development of the MTJ-based MRAM, an array of MTJs wired in the x- and y-directions was conceived as the densest memory, as illustrated in Fig. 5.2 (for example, see ref. [12]). The cell size of the cross-point cell is extremely small, thus it is very attractive for high-density cells.

It turns out that, circuit-wise, the MTJ behaves as a binary-state resistor, R_p, when the free layer and the pinned layer magnetizations are in parallel and as R_{ap} in antiparallel. Stray currents flow between the selected cell and unselected cells in the array. Thus, such a memory array is not practical; a gating element must be added to each MTJ to stop the stray current from flowing through the unselected cells in the array. The choices of gating element are a diode or a transistor.

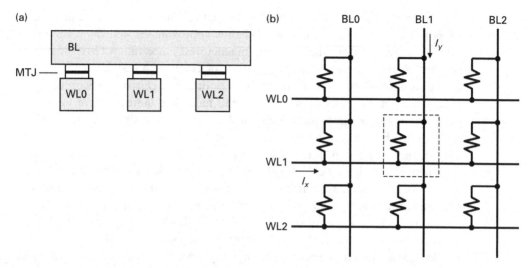

Figure 5.2. An array of MTJs, represented as resistors, connected by x- and y-direction wires forms a memory: (a) cross-section view of cells; (b) array circuit schematic. The selected cell is written by applying current in its bit line (BL) and its word line (WL). However, sensing the resistance of the MTJ is problematic due to stray currents flowing between the selected cell and unselected cells.

5.2.2   1T-1MTJ cell

A 1T-1MTJ (one magnetic tunnel junction, one transistor) cell is made up of a MTJ and a gating MOS transistor for the purposes of cell selection. Figure 5.3 shows the circuit schematic diagram and a cross-section of a typical field-MRAM cell. In each cell, one end of the MTJ is connected to the bit line (BL) and a MOSFET is on the other end. In addition, a write word line (WWL) is placed adjacent to the MTJ, in the orthogonal direction to the bit-line direction for writing. The gating transistor is controlled by a read world line (RWL).

To read a cell, the RWL of the selected cell is activated; the gating transistor is turned on, and a small current is applied to the selected BL that flows through the selected MTJ. The resistance of the MTJ is sensed to determine the data stored in the cell. The MOSFETs of the non-selected rows of cells are turned off during read, so that there is no stray current flowing from the selected cells to the unselected rows of MTJs.

To write the cell, currents are applied to both the WWL and the BL of a selected cell to generate hard- and easy-axis fields at the selected cell, respectively. The resultant field switches the free-layer magnetization based on the Astroid switching principle, and thus writes the cell. The direction of the BL current determines the direction of the easy-axis field, and thus data are written into the free layer of the MTJ. Note that all RWLs in the array are deactivated during the write operation, and there is no current flow in any MTJ.

This is the basis of a variety of the read/write operation schemes of field-MRAM cells. The read operation is common to all types of MRAM cells, namely detecting the binary resistance states of the MTJ.

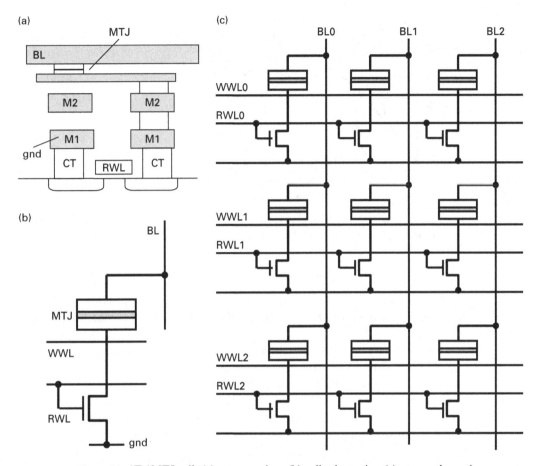

Figure 5.3. 1T-1MTJ cell: (a) cross-section; (b) cell schematics; (c) array schematics.

In the following, we will first discuss the read operation, and then we will discuss the principle of different field-write schemes.

5.3 Read signal

The maximum available read signal from a MTJ is the product of the voltage across the MTJ, V_{MTJ}, and the TMR ratio, i.e.

$$\Delta V_{\mathrm{read}} = V_{\mathrm{MTJ}} \times \frac{\delta R}{R_{\mathrm{p}}}. \tag{5.1}$$

Since $\frac{\delta R}{R_{\mathrm{p}}}$ falls slowly as V_{MTJ} increases, ΔV_{read} reaches a maximum and then decreases. An important MTJ figure of merit for the read operation is the product of the TMR ratio and V_{50}. The latter defines the voltage across MTJ at which $\frac{\delta R}{R_{\mathrm{p}}}$ falls to half the value at $V_{\mathrm{MTJ}} \approx 0$. Together they determine the voltage across

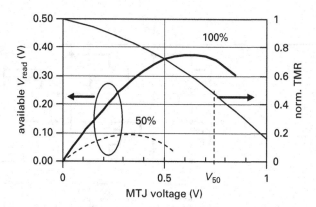

Figure 5.4. Normalized TMR vs. bias voltage, where V_{50} is the bias as the TMR decreases to half the zero-bias value. Available read signals are from MTJs with 50% and 100% TMR ratios.

the MTJ at which ΔV_{read} is maximum. Figure 5.4 illustrates V_{50} and the maximum available read signal for two TMR ratios. Note that the ΔV_{read} maximum occurs at voltages lower than V_{50}.

To sense a MTJ binary state, its resistance value is compared against that of a reference resistor. Typically, the reference resistor is set to the mid-point of the R_p and R_{ap}, or $0.5(R_p + R_{ap})$. The spread of the MTJ resistance in an array of MTJs determines the minimal TMR ratio required for sensing. In general, the R_p distribution is a Gaussian (or Normal) distribution, which can be characterized by a mean value and a sigma value, $\langle R_p \rangle$ and $\sigma(R_p)$, respectively:

$$n(R) = \frac{N}{\sqrt{2\pi}\sigma} \cdot \exp\left[-\frac{(R_p - \langle R_p \rangle)^2}{2\sigma(R_p)^2} \right], \tag{5.2}$$

where N is the total MTJ population in an array. The population of R that is between $\langle R_p \rangle \pm R$ equals $\int_{\langle R_p \rangle - R}^{\langle R_p \rangle + R} n(R)dR$. The integration of a Gaussian function results in an *error function*, $erf(x)$, where $x = \frac{R - \langle R_p \rangle}{\sqrt{2}\sigma(R_p)}$. The coverage between $\langle mean \rangle$ ± 1σ is 68.27% of the total populations; 2σ covers 95.45%; and 3σ covers 99.73%. The escape rate is $erfc(x) = 1 - erf(x)$, as illustrated in Fig. 5.5. To make a memory chip with more than one million bits, one needs to consider a 5σ distribution, so that the error probability is less than 1. Thus, the window of correct sensing is $TMR - n[\sigma(R_p)/R_p + \sigma(R_{ap})/R_{ap}]$, all units in %. Figure 5.6 illustrates the read margin.

5.3.1   Sense reference cell

The reference cell provides a reference voltage or current to the sense amplifier. Figure 5.7 illustrates a version of the reference cell in a read-sensing circuit.

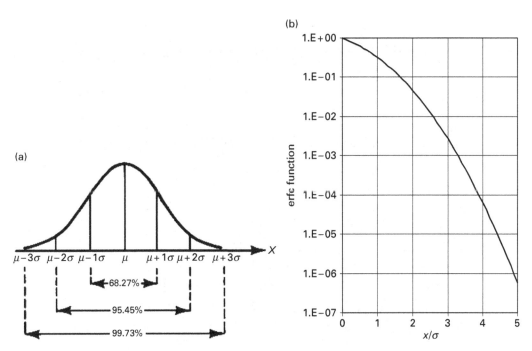

Figure 5.5. (a) Gaussian (Normal) distribution. (b) Escape rate in erfc function. At $\pm 5\sigma$, the escape rate is less than 10^{-6}.

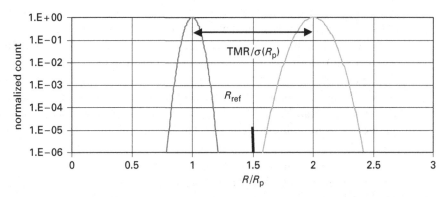

Figure 5.6. The read margin of a 1 Mb chip with TMR $= 100\%$ and $\sigma(R_p)/\langle R_p \rangle = 4\%$. At the 10^{-6} point, the read margin to the reference R_{ref} is nearly zero.

Since the MTJ is a non-linear diode with a voltage-dependent resistance, the reference cell is biased in a way identical to the basic cell in the memory array. Two basic cells are placed in shunt to form a reference cell: one stores "1" and the other stores "0". Thus, the sum of the MTJ currents is $(I_p + I_{ap})$, where I_p and I_{ap} are the currents through the MTJ in a parallel state and antiparallel state, respectively. This current is mirrored and then divided by 2 with four

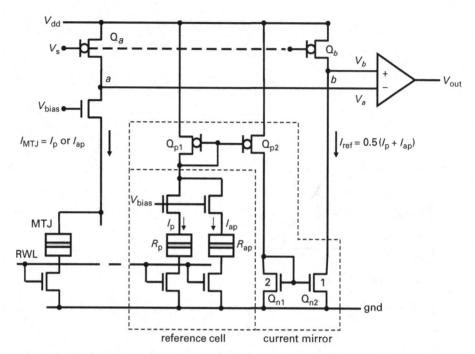

Figure 5.7. Reference cell and current mirror circuit to generate reference current for sense amplifier.

transistors, Q_{p1}, Q_{p2}, Q_{n1} and Q_{n2}, with the gate width ratio of Q_{n1} and Q_{n2} being 2:1 to form the reference current. Thus, $I_{ref} = 0.5(I_p + I_{ap})$.

A cell in a memory cell array is selected by activating the selected bit line and the selected read word line. On the bit line, transistor Q_{bias} on the bit line clamps the voltage of the selected cell and the reference cell, so that they are biased at the same voltage. The cell voltage is $V_{bias} - V_{gs}$, where V_{gs} is the gate-source voltage of the clamping transistor, which is operated over a linear region, such that it is insensitive to the transistor current. The cell current is either I_p or I_{ap}. Transistors Q_a and Q_b are operated in the near-saturation region, such that their drain voltages are sensitive to the drain current. Thus, a differential voltage $(V_a - V_b)$ is developed between node a and node b. The sense amplifier senses and amplifies the difference between V_a and V_b. When the cell current is I_p, which is smaller than I_{ref}, V_a is higher than V_b, and the sense amplifier output voltage, V_{out}, is positive. Otherwise, V_{out} is negative.

In this example, reference cells are placed in spare columns. The reference cells can also be placed in spare rows in the folded bit-line array architecture [13].

5.3.2   Sense amplifier

The sense amplifier amplifies the small signal developed between the bit line and the reference. Figure 5.8 shows an example of a sense-amplifier circuit.

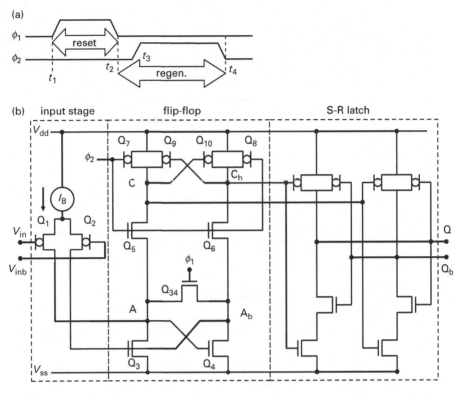

Figure 5.8. Sense-amplifier circuit. (a) The clock waveform controls the reset and signal regeneration time. (b) Three stages: input, flip-flop and S-R latch. (After ref. [14], with permission from the IEEE.)

The amplifier is a voltage comparator having an input stage that isolates the large capacitive loading of the bit lines from the amplifier internal circuit nodes. It is followed by a flip-flop and then a S-R (set-reset) latch. The dynamic operation of the amplifier is divided into two stages: a reset stage from time t_1 to t_2, and a regeneration stage from t_2 to t_4. During the reset stage, ϕ_1 is activated, the nodes A/A_b are shorted together by transistor Q_{34}, and the potentials on A and A_b are equalized. During this period, ϕ_2 is deactivated, and Q_5/Q_6 are off; Q_9 and Q_{10} are on, and nodes C/C_b are pre-charged to the same voltage, V_{dd}.

The regeneration starts after t_2 when ϕ_1 is deactivated prior to t_3, at which point ϕ_2 is activated; Q_{34} is off, and a differential voltage starts to develop at nodes A/A_b. Note that, since Q_5/Q_6 are off, nodes A/A_b are decoupled from the loading of the p-transistors (Q_7-Q_{10}) of the flip-flop, and thus the capacitance loading on nodes A/A_b is small. A differential voltage is developed quickly across the inputs of transistor Q_3 and Q_4. After t_3, clock ϕ_2 is activated, the source and drain of the Q_5/Q_6 pair are shorted, and Q_7/Q_8 are turned off. The differential voltage at nodes A/A_b is coupled up to the flop-flip through Q_5/Q_6 to nodes C/C_b, the gates of the cross-coupled pair, Q_9/Q_{10}. The flip-flop stage is set and a large differential voltage is developed, which resets the S-R latch.

The matching of the characteristics of comparator transistors Q_3 and Q_4 is very important. It determines the minimal voltage a sense amplifier can correctly detect. The transistor mismatch can be characterized by the standard deviation of its transconductance (g_m). To detect the two MTJ resistance states accurately, the TMR ratio must cover not only the variation of the R_p and TMR of the MTJs, σ_{MTJ}, but also the mismatch of comparator transistor pair, σ_{Tx}. In other words, the minimum of TMR must satisfy

$$\langle TMR \rangle > n\sigma,$$

where

$$\sigma = \sqrt{\sigma_{MTJ}^2 + \sigma_{Tx}^2}. \tag{5.3}$$

For an array of >1 Mb cells, counting 6σ from the spread of both "0" and "1", n should be greater than 12. Successful products are usually built with $n \sim 20$ [15].

5.4   Write bit cell with magnetic field

In the following sections, we will discuss the various design aspects of the field-write operation. First, we will discuss the conversion efficiency from the write current to the write field and the methods employed to improve the write efficiency. Then, we will discuss the write modes of the field MRAM. Historically, Astroid write mode was first developed for writing bit cells in a random access MRAM. This mode works well for single-free-layer MTJs in a straightforward manner. This write mode requires substantial engineering work to widen the operational margin. Subsequently, a toggle-mode write was developed on MTJs with synthetic antiferromagnetic free layers. Each has its pros and cons. There are other write modes which operate on different cell types and MTJ structures.

5.4.1   Write-field conversion efficiency

For the 1T-1MTJ cell structure shown in Fig. 5.3, the write field is generated by the current in both the write word line and in the bit line, which also serves a write bit line. Here we will discuss the write current efficiency, H/I, in the unit of Oe/mA. The bit line is in direct contact with the top electrode of the MTJ. The write word line is below the MTJ base plate and a dielectric insulator separates the two. For a long write line (compared to the MTJ dimension), one can estimate the write current efficiency based on the Biot–Savart Law (see Chapter 1). The write line is partitioned into many small cross-sectional elements; by summing the field vector over the contribution of each current element, one obtains the total field. The current that flows in the element that is closest to the MTJ free layer generates the field most efficiently. Since the write bit line is physically closer to the MTJ than the write word line, its field generation efficiency is larger. Figure 5.9 shows a

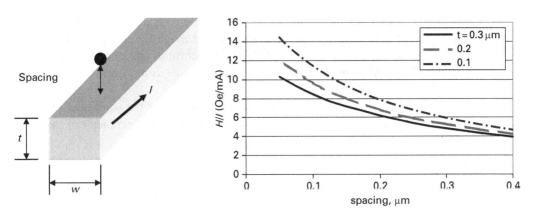

Figure 5.9. Write-field efficiency vs. spacing between write word line geometry and the spacing between the MTJ free layer and write line ($w = 0.25\,\mu m$).

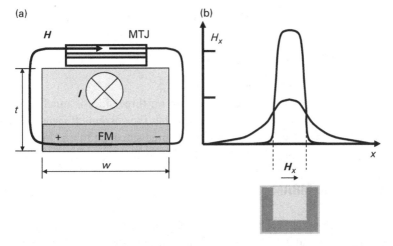

Figure 5.10. (a) Write line with ferromagnetic material one-sided cladding; (b) numerical simulation result of a wire with three-sided cladding.

calculated write current efficiency for wire width $= 0.25\,\mu m$. Maximum efficiency is achieved when the line height is reduced to the limit set by the electromigration and/ or excessive resistance in the write line. For this purpose, copper line is less resistive and more electromigration-tolerant than aluminum line, thus it is more efficient.

5.4.2 Write-line cladding

The write-current efficiency can be improved further by cladding a portion of the write line's surface with a soft ferromagnetic material. Figure 5.10 shows the cross-section of a write line with ferromagnetic cladding material at the bottom [16]. The easy axis of the FM cladding is along the length direction of the wire, and the magnetization of the FM cladding rotates toward the direction of the field generated by the write line. The demagnetizing field in the ferromagnetic layer partially cancels the H from the current, and boosts the magnetic field in the air. One may estimate the amount of boost using Ampère's Law. For a current I in the wire, Ampère's law yields $\oint H \cdot dl = I$.

The integration is carried out around the perimeter of the word line. Ignoring the corner effect, the field is approximately constant at the perimeter of the wire. Thus,

$$H \approx I/(2t + 2w), \tag{5.4}$$

where t and w are the wire width and height, respectively.

When the FM layer is inserted,

$$\oint H \cdot dl = \int\limits_{\text{air}} H_{\text{air}} \cdot dl + \int\limits_{\text{FM}} H_{\text{FM}} \cdot dl \approx \int\limits_{\text{air}} H_{\text{air}} \cdot dl, \tag{5.5}$$

since the demagnetizing field cancels most of the H inside the ferromagnetic cladding. Thus,

$$H_{\text{air}} \approx I/(2t + w), \tag{5.6}$$

i.e. the magnetic field is in air boosted.

Clearly, having ferromagnetic cladding on three sides is much more effective. Calculation and measurement both show that the current efficiency is increased in the range of 1.6–2.5x in a write line with three-sided cladding [16, 17].

5.5   Astroid-mode MRAM

Astroid-mode write is the appropriate choice for selective switching in a MTJ, in an array, with a single ferromagnetic free layer. For this write mode, the selected cell in an array is written with write currents in the selected write word line and the write bit line as illustrated in Fig. 5.11(a). In the early stages of MRAM development, this write mode suffers severe write disturbance problems, namely the data stored in a cell are frequently corrupted (or disturbed) when its neighboring cells are written. To understand why the disturbance happens, let us examine the switching-energy barrier of a cell when a field is applied to its neighboring cell. From there, we can discuss the engineering solution to the problems.

5.5.1   Switching-energy barrier of Astroid-mode write

We use Fig. 5.11(a) to illustrate the write action in an array. The cell labeled "S" is the selected cell to be written. A written current I_w is applied to its write word line and another, I_b, is applied to its write bit line. The cells on the write word line labeled "1" experience H_h (the write field in the hard-axis direction). Similarly,

(a)

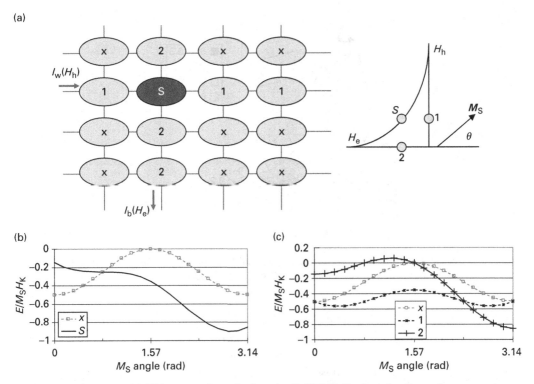

(b)   (c)

Figure 5.11. (a) Write operation of selected cell "S." Half-selected cells on the same write word line are marked "1" and those on the same write bit line are marked "2." Normalized energy barrier of (b) selected and un-selected cells, (c) half-selected cells. The applied field $H_e = H_h = 0.5^{3/2}H_K$.

the cells on the write bit line labeled "2" experience H_e (the write field in the easy-axis direction). These are the half-selected cells. The selected cell, labeled "S," experiences both H_h and H_e. The unselected cells, labeled "x," do not experience any field.

The write field of the selected and half-selected points on a switching Astroid plot is also shown on the right-hand side of Fig. 5.11(a). Suppose that the initial magnetization of the free layer points to zero degrees, or $\theta = 0$. After the write field increases, θ increases, and eventually M switches.

Figure 5.11(b) shows the normalized switching-energy barrier as a function of θ of the magnetization of the free layer of the unselected cells and the selected cell. The normalized energy barrier of the unselected cells is –0.5, while that of the selected cell is ~ 0, as it should be.

The switching-energy barrier of the MTJ on the half-selected bit line cells can be computed as follows. For the applied field in the x-direction, the normalized switching-energy barrier ε is given by

$$\varepsilon = E/H_K M_S = -0.5\cos^2\theta - h\cos\theta, \tag{5.7}$$

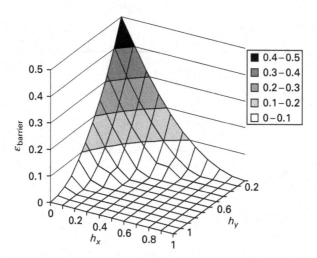

Figure 5.12. Cell energy barrier under external field in easy- and hard-axis directions below the switching threshold; $h_x = H_x/H_K$ and $h_y = H_y/H_K$.

where $h = H_x/H_K$. The energy minimum and maximum are at $d\varepsilon/d\theta = 0$. We obtain $\sin \theta = 0$ and $\cos \theta = -h$. There are two energy minima, at $\theta = 0$ or π and $\varepsilon_{\min} = -0.5 \pm h$. The positive sign applies to M_S in the opposite direction to the switching field H, and vice versa. The energy maximum $\varepsilon_{\max} = 0.5h^2$ Thus, the energy barrier is given by

$$\varepsilon_{\text{barrier}}(h) = \varepsilon_{\max} - \varepsilon_{\min} = 0.5h^2 - (-0.5 + h) = 0.5(1 - h)^2. \tag{5.8}$$

The energy barrier of the half-selected cells along the direction of the bit line is reduced by a factor of $(1 - h)^2$. It is illustrated in Fig. 5.11(c).

Similarly, the switching-energy barrier of the half-selected cells on the write word line is reduced by the same amount. However, there is a difference between the two kinds of half-selected cells. For those on the same bit line, the energy level of the two energy minima is different. This implies that there is a finite probability that the free layer attempts to switch from the higher-energy minimum to the lower one. In other words, there is a tendency that the M_S switches from its initial value $\theta = 0$ to a new value $\theta = \pi$. For those cells on the same word line, the energy minima are the same. There is no preferred direction of switching.

The switching-energy barrier of a cell under both easy-axis and hard-axis fields is shown in Fig. 5.12. The constant-energy contours are a set of curves in parallel to the switching Astroid curve. The switching Astroid represents the zero-switching-energy barrier. Since the energy barrier of the half-selected cells is much smaller than those that are in idling mode, these half-selected cells are more vulnerable to write disturbance (i.e. an unwanted state change) than the idling cells.

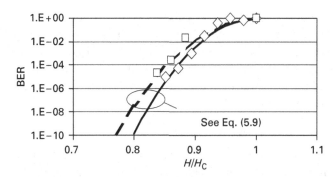

Figure 5.13. Measured and fitted switching probability of sub-threshold field write at 25 °C (diamond, solid line) and at 85 °C (square, dashed line); H_c is the switching threshold field in the easy-axis direction. (After ref. [19].)

5.5.2   Write-error rate of a bit cell

Micromagnet simulations [18] predict the half-selected bit error rate (BER) to be

$$\text{BER} \sim \exp[-A(1 - H/H_K)^2/k_B T], \tag{5.9}$$

where A is proportional to the switching-energy barrier. It is experimentally verified as shown in Fig. 5.13. It shows the switching probability of a single cell under a field below the switching threshold at two different temperatures. This is a temporal statistics of the write-bit-error rate, a kind of soft error. It means the switching is probabilistic, not deterministic. The definition of "switching threshold" is merely a reference.

In a MRAM array, the actual field applied to the selected cells is much larger than the threshold defined by the switching Astroid to ensure that the cell is written at a low write-fail rate. As a result, the half-selected cells are disturbed more severely or more often. This problem has been plaguing MRAM designers for a long time. The solution comes from adjusting the MTJ properties such that the MTJ exhibits different switching characteristics.

5.5.3   Write soft error rate of an array of memory cells

As shown in Chapter 3, the probability of switching, $p(t)$, is an exponential function of the energy barrier:

$$p(t) = f \exp(-E_b/k_B T) < 1, \tag{5.10}$$

where f is related to the gyro-magnetic spin precession frequency and is of the order of 10^9 per second. For a single MTJ to retain data for 10 years (3.15×10^8 seconds), $E_b/k_B T > \ln (10^9 \times 3.15 \times 10^8) \sim 40$. For a 128 Mb memory chip with a 128 Mb MTJ to keep data over 10 years, $E_b/k_B T > \ln (4.23 \times 10^{25}) \sim 59$. With such an energy barrier between the two MTJ states, the probability of any single bit failure is less than 1 over 10 years. This projection is correct if the chip is in stand-by mode or in storage, rather than being written.

When the chip is in operation, it is inevitable that a fraction of the cells in a memory array exist under the half-selected condition. Since the energy barrier of the half-selected cells is lower than that of the unselected cells, one may calculate the required energy barrier to be

$$P_{total} = N_{idling} \times p_{idling}(t) \times t_{idling} + N_{hs} \times p_{hs}(t) \times t_{hs} < 1, \qquad (5.11)$$

where P_{total} is the sum of the unwanted switching probabilities of N_{idling} stand-by cells over a time period t_{idling} and N_{hs} half-selected cells over time t_{hs}. Note that the energy barrier of the half-selected cells is smaller than that of the stand-by (and idling) cells, thus $p_{idling}(t)$ is much smaller than $p_{hs}(t)$.

Homework

Q5.1 Consider a 1 Mb bit MTJ memory in a 1024×1024 1T-1MTJ cell array. The length of the write word line extends across the array in one direction and so does the write bit line. Thus, during the write operation, there are 2048–1 half-selected cells. The chip is constantly being operated, with one-fifth of the time in write mode and four-fifths of the time in read mode. Calculate the energy barrier required to operate the chip for 10 years without write disturbance failure.

A5.1 The probability of bit fail in a stand-by state is p_{st} and that in a half-selected state is p_{hs}:

$$p(t) = [(1024 \times 1024) - (2047)]p_{st}\ t_{st} + (2047)p_{hs}\ t_{hs},$$

where

$$t_{st} = 3.15 \times 10^8 \text{ s and } t_{hs} = 0.2t_{st},$$

$$p_{st} = \exp(-E_b/k_BT) \text{ and } p_{st} = \exp(-(0.4E_b)/k_BT).$$

The second term dominates, since the energy barrier is much smaller in the half-selected cells, so they are more vulnerable. For this case, $p(t) \approx 2047 \times p_{hs} \times 6.31 \times 10^7 = 1$. Thus $E_b/k_BT = 64$.

5.5.4   Solution to the write disturbance problem

The write field of the Astroid-write mode is bounded. For a given hard-axis write field, the upper bound of the easy-axis write field is the half-selected disturbance, and the lower bound is the Astroid. When thermally activated magnetic reversal is taken into consideration, the safe operating field regime is small [20]. In addition, variability in the write threshold can result from incoherent magnetization rotation in the MTJ free layer [21–23]. The write-field operation regime

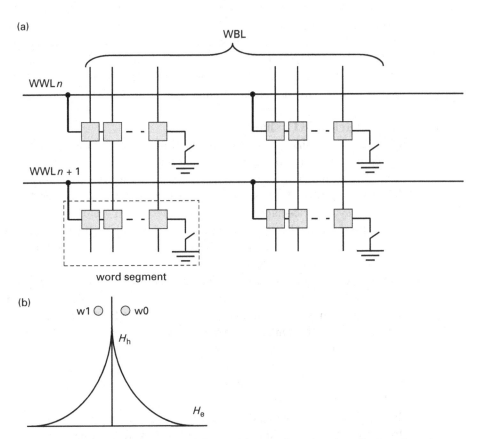

Figure 5.14. (a) Segmented write word line architecture introduced to alleviate the write disturbance problem of a field MRAM. Each box is a 1T-1MTJ cell; WWL and WBL denote write word line and write bit line, respectively. (b) Write operation on Astroid: write the entire segment of cells with a very large H_h and small $\pm H_e$ to alleviate the disturbance of the half-selected cells on bit lines. The Astroid is shown as a reference for the operating points. (After ref. [25].)

is further reduced. There are many ways to solve the write error rate [24]. Here, we will give two examples: one is to change the circuit architecture of the cell array, while the other is to change the MTJ switching-energy-barrier characteristics. Both improve the write margin of the selected and half-selected cells.

5.5.4.1 Segmented write

One way in which to alleviate the disturbance to the half-selected cells is to construct the cell array into segments in one of the word or bit directions. Figure 5.14 illustrates a segmented-word cell array architecture [25]. The write word line is divided into many segments. The write current in each segment is controlled by a bit direction control. During read or write, all cells in the selected segment are written and read simultaneously. In this manner, there is no write disturbance problem in the word direction, since the cells in the whole segment are written. The write disturbance problem in the bit line direction is alleviated by increasing the write-word-line current and reducing the write-bit-line current. This is illustrated in Fig. 5.12. The probability of causing disturbance to the cells on the half-selected bit line is minimized.

5.5.4.2   Asymmetrical MTJ shape

Many different shapes of MTJ were proposed to reduce the variation of the switching field and to enlarge the margin between the switching field of the full-selected and half-selected cells, thus improving the selectivity of the field-MRAM cell array. A key solution is to use an asymmetric-shaped MTJ [26–28].

In an asymmetric-shaped MTJ, the magnetization vectors at the two long ends are trapped by the demagnetizing field and the cell is in the C-state. Having only one remnant state, rather than both the S-state and the C-state, the spread of the switching threshold is reduced (see Section 3.3.3).

The free-layer magnetization switching behavior is not uniform, nor coherent. The switching behaviors at the ends of the free layer are different from those in the middle. Thus, the switching properties cannot be modeled using a coherent rotation model, and micromagnetic simulation is required to take into account these different local behaviors. The shape anisotropy of the asymmetrical shape is balanced by the exchange energy of the free layer and results in a very large switching threshold margin between the selected and half-selected cells.

The selected cell experiences both easy- and hard-axis fields. As shown in Fig. 5.15, the hard-axis field is first applied in a direction that turns the free layer from a C-state to an S-like state (Fig. 5.15(2)), before the easy-axis field is turned on. Then, the easy-axis field is applied to switch a selected cell in an S-state (Fig. 5.15(3)). After the switching, the fields are removed, and the cell returns to a remnant C-state (Fig. 5.15(5)).

On the other hand, only one of the two write fields acts on the half-selected cells (either the easy-axis or the hard-axis field). For the cells that experience only the hard-axis field, their magnetization vectors at the two ends of the MTJ rotate and shift from a C-state towards S-state, similar to the selected cell (as shown in Fig. 5.15(2)). Without the easy-axis field, the cells remain in the S-state and do not switch. Those that experience only the easy-axis field, as illustrated in Fig. 5.16(2), tend to bend the C-state further and form a magnetic vortex in the middle of the free layer, making the MTJ harder to switch. The easy-axis field required to switch a cell in the C-state must be strong enough to annihilate the vortex. This field is much larger than the switching threshold of a cell in the S-state. Thus, the asymmetric MTJ design enlarges the switching threshold margin between the half-selected cells and the full-selected cell. From the level of switching current, I_{BL} of Figs. 5.15(3) and Fig. 5.16(3), one finds that the margin can be as large as $\sim 6\,\text{mA}$ ($8.2 - 2.2\,\text{mA}$).

Quantitatively, the margin can be adjusted by modifying the shape of the asymmetrical MTJ. Details can be found in refs. [27] and [29]–[31].

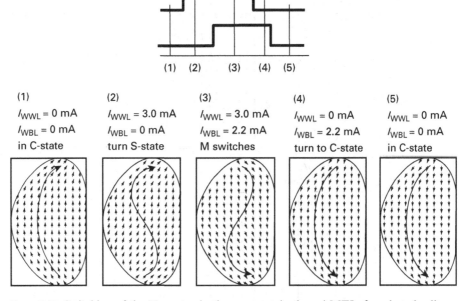

Figure 5.15. Switching of the M_S vector in the asymmetric-shaped MTJ of a selected cell. The hard-axis field turns the M_S vector from a C-state to an S-state prior to switching. (After ref. [26].)

5.6   Toggle-mode MRAM

A toggle-mode MRAM patent was issued to L. Savtchenko in 2003 [32]. It improves the write margin between the selected cell and the half-selected cells, and thus overcomes some of the weaknesses of the Astroid-mode write.

5.6.1   Toggle-mode cell

Unlike in the Astroid-mode MRAM, the MTJ long axis (easy axis) of the toggle-mode MRAM is oriented at 45 degrees from the word line–bit line grid, as shown in Fig. 5.17. The free layer of the MTJ is made up of a synthetic antiferromagnetic (SAF) free layer. Typically, the non-magnetic layer between the two ferromagnetic layers is ruthenium, which provides a strong RKKY interlayer coupling (see Chapter 2). In this case, the thickness of ruthenium is in the range of 7–12 Å (first cycle) and 18–23 Å (second cycle). The interlayer coupling is in the antiparallel state and $J = 0.044$ erg/cm^2 in the first cycle and is weaker in the second cycle for CoFeB/Ru/CoFeB SAF films.

There is no net magnetic moment in the SAF free layer when the moment of the SAF is balanced, i.e. $M_1 t_1 - M_2 t_2 \approx 0$, where M is the magnetization and t is the thickness of each ferromagnetic layer, and the subscripts 1 and 2 denote the two ferromagnetic layers of the SAF film stack. As described in Chapter 3, when the write field is smaller than the spin-flop field, H_{SF}, the SAF free layer

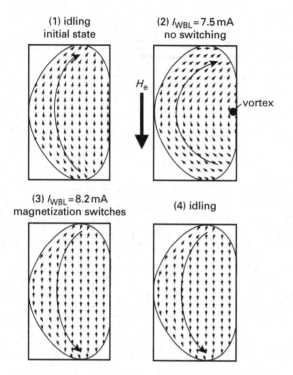

Figure 5.16. Magnetization vectors in asymmetric-shaped MTJ of a half-selected cell. (1) MTJ in idling state; (2) without the hard-axis field, the easy field forces the magnetization in the free layer to form a vortex that increases the switching threshold; (3) at a higher easy-axis field, large enough to annihilate the vortex, the magnetization switches; (4) idling state after switching.

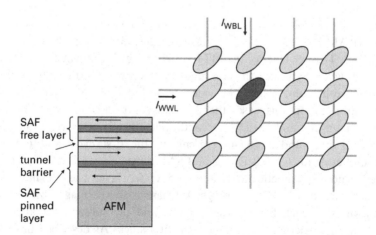

Figure 5.17. The MTJ film stack of a toggle-MRAM and the layout of the cell array.

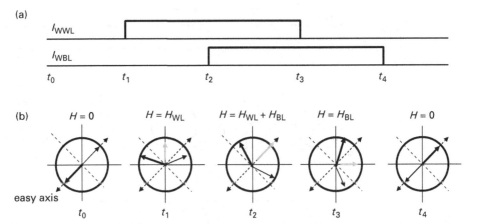

Figure 5.18. The timing sequence of the write field pulse and behavior of the magnetization in a SAF free layer.

does not react to the write field. When the write field is greater than H_{SF}, the two magnetizations of the SAF free layer "scissor," or tilt, toward the direction of the field and deviate from the antiparallel state. Eventually, as the write field becomes greater than the saturation field, H_{ssat}, the two free layers align to the write field.

5.6.2   Switching of SAF free layer in toggle-mode write

The timing sequence of the applied field to the selected cell is shown in Fig. 5.18(a). The word-line write current is applied first, then a bit-line write current is applied. The two are applied with a phase delay. The behaviors of the two SAF free layers under different phases of the toggle write are shown in Fig. 5.18(b). During idling ($t < t_1$), M_1 and M_2 of the SAF free layer are in an antiparallel state, and are in the long-axis direction. At $t_1 < t < t_2$, only a word-line field, greater in value than H_{SF}, is applied at 90°, and this field induces spin-flop. Now, M_1 and M_2 rotate from the idling direction to a new direction. At this step, the magnetization M of one of the two free layers rotates by an angle $>45°$. The SAF free layer exhibits a net moment pointing in same direction as the field direction. During the time interval $t_2 < t < t_3$, both the word-line field and the bit-line field are applied. The resultant field rotates clockwise from 90° to 45°, and so does the net moment of the SAF free layer. During $t_3 < t < t_4$, the word-line field is off, so only the bit-line field remains, which is in a $-45°$ direction. The net moment of the SAF free layer completes a 90° rotation. At the end of this period, each M of the SAF free layer has rotated over 135° from its initial position. After t_4, the bit-line field is removed, and each M of the SAF free layer relaxes to the long-axis direction, which is less than 45° away. Thus, each toggle operation rotates the free layer through 180° from their initial positions.

Such toggle action flips the SAF free layer, independent of the previous state. Thus, it toggles the resistance of the MTJ and the data previously stored in the

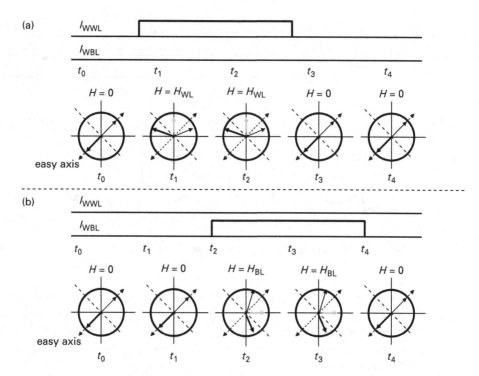

Figure 5.19. Magnetization behavior of half-selected cells under (a) the word-line field only and (b) the bit-line field only.

cell. The direction of the write field can be restricted to one polarity. It is different from the bit-line write field of the Astroid-mode MRAM. Therefore, to write data into a cell, one must read out the data that are originally in the cell prior to writing and compare the two. If they differ, the toggle action is activated to flip the cell data. If they are the same, toggle action is not activated.

The half-selected cells on the same write word line or write bit line experience only one field. A single write field induces spin-flop in the SAF free layer. Both M_1 and M_2 remain in the same quadrant of their initial direction. When the write field is removed, M_1 and M_2 relax back to their initial states. Thus, toggling does not happen and the half-selected cells do not switch. Figure 5.19 illustrates magnetization behavior under (a) the word-line field and (b) the bit-line field.

The trajectory in the phase diagram of the write-field vector on the free layer is shown Fig. 5.20. The selected cell sees a complete field locus while the half-selected cells see only a segment of the trajectory.

5.6.3   Energy diagram of toggle operation

To understand the switching properties of the toggle-mode MRAM, one must start from the construction of an energy diagram as a function of the applied fields. As shown in Chapter 3, the energy of each ferromagnetic layer of a

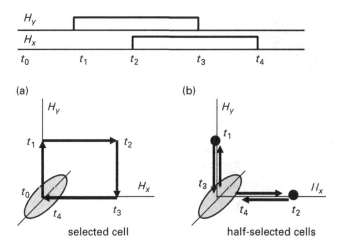

Figure 5.20. Trajectory of write field on a selected cell (a) and half-selected cells (b).

balanced SAF free layer consists of anisotropy energy, magnetostatic energy and RKKY interlayer coupling energy. The MTJ shape anisotropy is absorbed into H_K for simplicity. Here, we ignore the interlayer (Néel) coupling due to correlated surface roughness and the stray field from the fixed layer, which biases the free layers. From Chapter 3, the energy per unit area of the SAF film stack is given by

$$E = -0.5 H_K (M_1 t_1 \cos^2 \theta_1 + M_2 t_2 \cos^2 \theta_2) + J_{RKKY} \cos(\theta_1 - \theta_2) \\ - H(M_1 t_1 \cos(\theta_1 - \phi) + M_2 t_2 \cos(\theta_2 - \phi)) \tag{5.12}$$

We can simplify the analysis further by considering the case $M_1 t_1 = M_2 t_2 = Mt$, $j = J_{RKKY}/(H_K Mt)$ and $h = H/H_K$. Then, the normalized Eq. (5.14) becomes

$$\varepsilon = E/(H_K Mt) = -0.5 \cdot (\cos^2 \theta_1 + \cos^2 \theta_2) + (j) \cdot \cos(\theta_1 - \theta_2) \\ - (h) \cdot [\cos(\theta_1 - \phi) + \cdot \cos(\theta_2 - \phi)]. \tag{5.13}$$

This equation allows one to find the angle of the magnetization vectors of the SAF free layers under the applied field. When the applied field is absent, there are two energy minima, as illustrated in Fig. 5.21, which exhibits a symmetry centered along a diagonal line. The angles of the two lowest-energy points, A and C (θ_1, θ_2), are at $(0, \pi)$ or $(\pi, 0)$, respectively; the magnetizations of the SAF free layer are pointing along the long side of the MTJ and are antiparallel. The maximum-energy points are at $(\pi/2, \pi/2)$ and $(-\pi/2, -\pi/2)$, and both magnetizations are pointing to the short side of the MTJ and are parallel. The energy barrier is the lowest-energy "hill," or a local maximum, point B, in the (θ_1, θ_2) diagram between these two energy minima (A and C), which is located at the mid-point between the two energy minima.

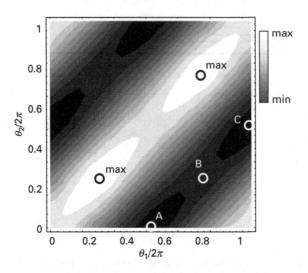

Figure 5.21. Energy contour of the SAF free layer in the absence of an applied field. Energy minima (A, C), a local "hill" (B) and energy-maxima points "max" are marked.

When a first write current pulse is applied, the field is at angle of $\pi/4$ from the long axis of the MTJ and its amplitude is large enough to induce spin-flop in the SAF free layer. The energy minima shift to new positions (see Fig. 5.22). The energy barrier between the local minima can then be located on the energy diagram, as illustrated in Fig. 5.22. Compared to Fig. 5.21, the energy minima shift to points A and C. The two magnetizations of the SAF free layers are no longer antiparallel, indicating that the magnetization of the SAF free layer scissors. The energy barrier between A and C is at point B, which also changes from the zero-field case.

As the total write field rotates clockwise, the energy minima (A and C) follow the field to new angles. Comparing the minima in Fig. 5.23 to Fig. 5.22, the angle shift is clear. In other words, the pair of magnetization vectors M_1 and M_2 rotate with the field. If the energy minima rotate past an angle of $\pi/2$ from their initial positions, the magnetization of the SAF will not relax back to its initial angle after the field is removed. Rather, the magnetization of the SAF switches polarity.

Note that the direction of the field acting on the half-selected cells does not rotate; it stays at an angle $\pi/4$ from the long axis of the MTJ. The energy minima move from the positions shown in Fig. 5.22 to those shown in Fig. 5.23 and return to their initial positions after the field is removed. Thus, the energy minima in the half-selected cells do not move any further.

The switching-energy barrier at point B in Figs. 5.22 and 5.23 can be calculated as a function of the applied field. For the write field in the direction of the long axis, the switching-energy barrier shrinks to a minimum at $H = H_{SF}$, the spin-flop field. However, the switching-energy barrier increases when H along angle $\pi/4$

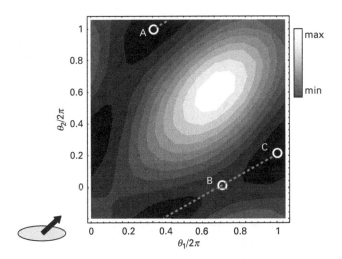

Figure 5.22. Energy contour plot of the magnetization angles of SAF free layers. The applied field is in the direction of $\pi/4$ from the long axis of the MTJ, as shown in the lower left corner. This field direction applies to both selected and half-selected cells. The energy-minimum point C does not pass the energy "hill" in Fig. 5.21.

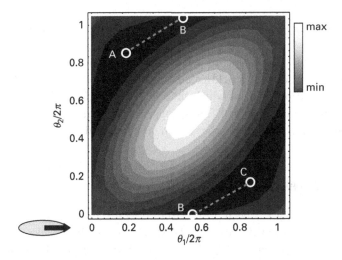

Figure 5.23. Energy contour plot of the magnetization angles of SAF free layers. The applied field is in the direction of the long axis. This field direction applies only to the selected cells. The energy-minimum point C now passes the energy "hill" B in Fig. 5.22, and thus the selected cell switches.

increases. This explains why the half-selected cells in the toggle-mode write array are more resistant to the write disturbance [20, 33]. The switching-energy barrier is illustrated in Fig. 5.24.

When the two FM layers of the SAF free layer are perfectly matched, the free layer switches in toggle mode. Note that, unlike the Astroid-mode switching, the

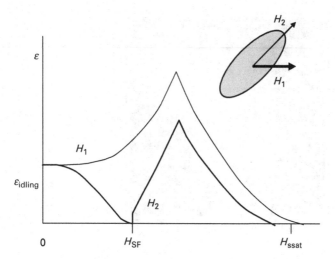

Figure 5.24. Normalized energy barrier as a function of applied fields along the easy-axis direction (H_2) and $\pi/4$ from the easy axis (H_1); H_1 corresponds to the field applied to the half-selected cells and H_2 corresponds to the full-selected cell. (After ref. [33], with permission from the IEEE.)

toggle-switching region does not intercept with either axis. The margins of write disturbance for the half-selected cells are larger than those in an idling state.

Figure 5.25 shows the toggle-mode operation regions in the field diagram. The operation zone starts at a field greater than H_{SF}, the spin-flop field, and ends at H_{ssat}, the saturation field, at which time the two magnetization vectors point in the same direction as the field. Thus, toggle-mode switching is more immune to the half-selected cell disturbance. When not matched, the MTJ switches in direct mode.

5.6.4   Write-current reduction

Toggle-mode switching has been proven to be effective in solving the disturbance problem of the half-selected cells in MRAM. One of its drawbacks, however, is that the switching current is very high. At the 180 nm node, the write current is greater than 10 mA. Many methods are proposed to lower the write current, but we will not discuss these in this book. Interested readers are referred to refs. [29] and [33]–[40].

5.7   Characterization method of MRAM chip write performance

The full-selected and half-selected cell performance figures are two of the most important characteristics of the write performance of MTJs in a memory array. The full-selected cell performance provides the distribution of the write threshold

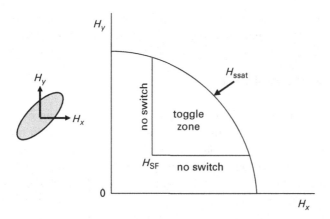

Figure 5.25. Toggle-mode operation zones; H_{SF} is the spin-flop field and H_{ssat} is the saturation field.

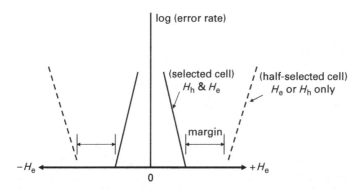

Figure 5.26. The full-selected cell error rate (solid lines), the half-selected cell error rate (dashed lines) and the write margin of a chip with segmented word-line architecture. The half-selected cells are on the same bit line as the selected cell.

field (or currents) of the MTJ over the entire array. The half-selected cell perform-ance is a measure of the disturbance rate of the half-selected cells as a function of the write field (or current). The chip operation window (illustrated in Figure 5.26) denotes the margin between these two parameters. Under a constant write field in the hard-axis direction, more MTJs in the memory array begin to switch as the easy-axis field increases, and the write fail rate drops. This is shown as the "selected cell" curve. The selected cell curve is measured with both the hard-axis (H_h) and easy-axis (H_e) fields. The slope of the selected cell curve (solid line) is the standard deviation of the σ of switching threshold.

As the write field continues to increase, some of the half-selected cells are disturbed, and the error gradually increases. Similarly, the slope of the half-selected cell curve error rate count (dashed line) is the distribution of the half-selected cell disturbance. The half-selected cell curve is measured with the easy-axis field (H_e) only. The chip operating margin is the space between these two curves.

5.8   Thermally assisted field write

This write method relies on the combination of thermal heating of the storage layer and the magnetic field in order to switch the magnetization direction, as illustrated in Fig. 5.27. The free layer of the MTJ consists of a ferromagnetic electrode and an antiferromagnetic layer with low blocking temperature (T_b) (see Chapter 4). During write, a current is applied to the MTJ, which heats up the MTJ above the blocking temperature. Thus, the free layer is no longer pinned and becomes free. Consequently, it can be switched by the write field. After switching, the current through the MTJ is removed, and, as the antiferromagnetic layer cools down below the blocking tempearture, it pins the free layer in a new direction. The fixing of the free layer improves the thermal stability of the cell.

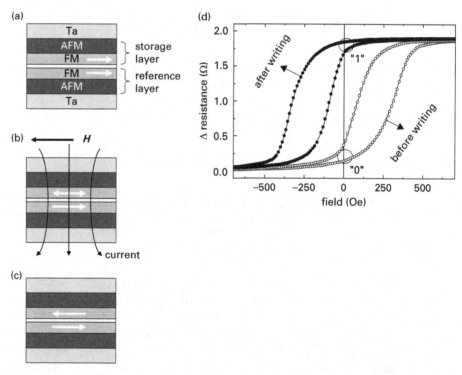

Figure 5.27. (a) Film stack of MTJ for thermal switching. The free layer consists of a soft FM layer (NiFe) and an AFM (IrMn) layer. (b) When switching the free layer with an external field, a current is applied to the MTJ to heat up the AFM beyond the blocking temperature. (c) After the heating current and the write field are removed, the AFM pins the free layer again, in the new free-layer direction. (d) The R–H loop (resistance of MTJ vs. applied field in the easy-axis direction) of the free layer before and after the switching. (After ref. [41], with permission from the IEEE.)

5.9   Multi-transistor cells for high-speed MRAM operation

For high-speed MRAM operation, a one magnetic tunneling junction, two transistor (2T-1MTJ) cell structure and a two magnetic tunneling junction, five transistor (5T-2MTJ) cell structure were proposed [16, 42]. Rather than writing data into a cell with currents in two orthogonal wires, the cell is written with one current, which is fed directly into the lower electrode via two gating transistors in the cell. No other cells in the memory array experience the field from this write current during the write operation, thus there is no half-selected cell disturbance issue. One can write with a current much greater than the write threshold. Thus, the switching speed can be very fast [43]. The drawback is that the two transistors in the cell are very large so that they can switch the write current on and off. Thus the cell size is large. In the Astroid-write and toggle-write schemes, the transistor in the cell switches only the read current, which is typically in the order of tens of microampères, two orders smaller than the write current. Figure 5.28(a) illustrates the field generation scheme of the 2T-1MTJ cell. The write current flows through the lower electrode of the MTJ, very close to the free layer, and thus the current-field conversion efficiency is very high. The field is in a direction 45° from the MTJ

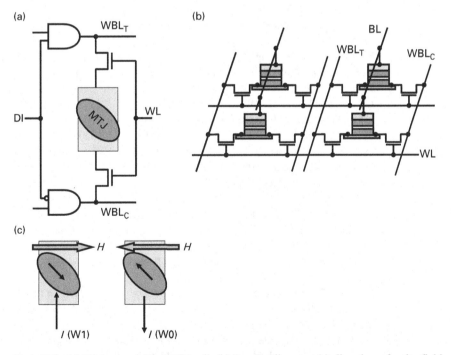

Figure 5.28. (a) High-speed 2T-1MTJ cell; (b) 2 × 2 cell array; (c) direction of write field. Only the selected cell receives the write current; there is no upper bound to the write current from the perspective of the half-selected cell disturbance. (After ref. [42], with permission from the IEEE.)

long axis, as illustrated in Fig. 5.28(c). The write current is of the order of few milliampères. A third terminal is connected to the top electrode of the MTJ for the read operation. Figure 5.28(b) illustrates a 2×2 cell array. The read scheme is the same as for other cells. Experimentally, the 1MTJ-2T cell structure accelerates the operating speed to 200 MHz.

One may further shorten the read access time by amplifying the read signal within the cell prior to feeding the signal to the bit line, rather than waiting for the signal to develop in the bit line. Since many cells are connected to the bit line, the bit line is loaded heavily and acts like a large capacitor. The large sense signal on the bit line allows the sense amplifier to latch earlier. By adding three more transistors and one more MTJ in each cell, a 5T-2MTJ cell amplifies the read signal from a few tens of millivolts to the full V_{dd} level at the bit line. Having such a large differential signal on the pair of bit lines of each cell, the sense amplifier can be operated over nanosecond time scales and a read cycle over 500 MHz has been demonstrated [42].

References

[1] A. V. Pohm, J. M. Daughton, J. Brown and R. Beech, *IEEE Trans. Magnetics* **31** (6, Pt. 1), 3200 (1995).

[2] J. -C. Wu, H. L. Stadler and R. R. Katti, *High Speed Magneto-resistive Random Access Memory*, US patent 5173873 (1992).

[3] K. T. Kung, D. D. Tang and P. -K. Wang *et al.*, *Nonvolatile Magnetoresistive Storage Device Using Spin Valve Effect*, US patent 5343422 (1994).

[4] T. Inoue, *Superconductor Magnetic Memory Cell and Method for Accessing the Same*, US patent 5276639 (1994).

[5] J. M. Daughton, "Magnetic spin devices: 7 years from lab to product," Symposium X, MRS Fall Meeting, Boston, MA, Dec. 1 (2004).

[6] D. D. Tang, P. K. Wang, V. S. Speriosu, S. Le, R. E. Fontana and S. Rishton, *IEDM Technical Digest* (1995), p. 997.

[7] J. S. Moodera, R. Meservey, and X. Hao, *Phys. Rev. Lett.* **70**(6), 853 (1993).

[8] T. Yaoi, S. Ishio and T. Miyazaki, *J. Magn. & Magn. Mater.* **126**, 430 (1993).

[9] R. Meservey and P. M. Tedrow, *Phys. Rep.* **238**(4), 214 (1994).

[10] K. Matsuyama, H. Asada, S. Ikeda and K.Taniguchi, *IEEE Trans. Magnetics*, **31**(6), 3176 (1995).

[11] W. J. Gallagher and S. S. P. Parkin, *IBM J. Research & Development, Special issue on Spintronics* **50**(1) (2006).

[12] Y. Asao, T. Kajiyama, Y. Fukuzumi *et al.*, *IEDM Technical Digest* (2004).

[13] T. Tsuji, H. Tanizaki, M. Ishikawa *et al.*, *Symposium on VLSI Circuits, Digest of Papers*, Honolulu, June 17 –19 (2004), p. 451.

[14] G. M. Yin, F. O. Eynde and W. Sansen, *IEEE J. Solid-State Circuits* **27**(2), 208 (1992).

[15] S. Tehrani, B. Engel, J. M. Slaughter *et al.*, *ISSCC Technical Digest Papers* **43**, 130 (2000).

[16] H. Honjo, R. Nebashi, T. Suzuki, S. Fukami, N. Ishiwata,T. Sugibayashi and N. Kasai, *J. Appl. Phys.* **103**, 07A711 (2008).

[17] K. S. Kim, C. E. Lee and S. H. Lim, *IEEE Trans. Magnetics* **39**(5), 2857 (2003).

[18] J. Zhu and X. Zhu, Private communications (2001).

[19] M. Lin, Private communication (2004).

[20] T. Yamamoto, H. Kano, Y. Higo *et al.*, *J. Appl. Phys.* **97**, 10P503 (2005).

[21] A. F. Popcov, L. L. Savchenko, N. V. Vorotnikov, S. Tehrani and J. Shi, *Appl. Phys. Lett.* **77**, 277 (2000).

[22] J. Shi and S. Tehrani, *Appl. Phys. Lett.* **77**, 1692 (2000).

[23] N. Shimonura, T. Kishi, M. Yoshikawa, E. Kitagawa, Y. Asao, H. Hada, H. Yoda and S. Tahara, *IEEE Trans. Magnetics* **41**(10), 2652 (2005).

[24] H. Hönigschmid, P. Beer, A. Bette *et al.*, *IEEE Digest ISSCC* (2006), paper 7.3, p. 136.

[25] W. R. Reohr and R. E. Scheuerlein, *Segmented write line architecture for writing magnetic random access memories*, US Patent 6335890 (January 1, 2002).

[26] K. Ounadjala and F. B. Jenne, *Asymmetric dot shape for increasing select-unselect margin in MRAM devices*, US Patent 6798691 (2001).

[27] T. Kai, M. Yoshikawa, Y. Fukuzumi *et al.*, *IEDM Technical Digest* (2004), p. 583.

[28] M. Nakayama, T. Kai, S. Ikegawa *et al.*, *IEEE Trans. Magnetics* **42**(10), 2757 (2006).

[29] S. C. Oh, J. E. Lee, H. -J. Kim *et al.*, *J. Appl. Phys.* **97**, 10P509 (2005).

[30] Y. Iwata, K. Tsuchida, T. Inaba *et al.*, *IEEE Digest ISSCC* (2006), paper 7.4, p. 138.

[31] T. Takenaga, T. Kuroiwa, J. Tsuchimoto, R. Matsuda, S. Ueno, H. Takada, Y. Abe and Y. Tokuda, *IEEE Trans. Magnetics* **43**(6), 2352 (2007).

[32] L. Savtchenko, B. N. Engel, N. D. Rizzo, M. F. Deherrera and J. A. Janesky, *Method of Writing to Scalable Magnetoresistive Random Access Memory Element*, US patent 6525906 (2003).

[33] D. C. Worledge, *Appl. Phys. Lett.* **84**, 2847 (2004).

[34] M. Motoyoshi, I. Yamamura, W. Ohtsuka *et al.*, *IEEE Symposium on VLSI Technology*, Honolulu, June 15–17 (2004), p. 22.

[35] T. Suzuki, Y. Fukumoto, K. Mori *et al.*, *IEEE Symposium on VLSI Technology, Technical Digest*, Kyoto, June 14–16 (2005), papers 10B-3, p. 188.

[36] Y. Fukumoto, T. Suzuki and S. Tahara, *Appl. Phys. Lett.* **89**, 061909 (2006).

[37] C. -C. Hung, Y. -J. Lee, M. -J. Kao *et al.*, *Appl. Phys. Lett.* **88**, 112501 (2006).

[38] D. W. Abraham and D. C. Worledge, *Appl. Phys. Lett.* **88**, 262505 (2006).

[39] S. -Y. Wang, H. Fujiwara and M. Sun, *J. Appl. Phys.* **99**, 08N903 (2006).

[40] Y. -J. Lee, C. -C. Hung, D. -Y. Wang *et al.*, *Appl. Phys. Lett.* **90**, 032503 (2007).

[41] I. L. Prejbeanu, W. Kula, K. Ounadjela, R. C. Sousa, O. Redon, B. Dieny and J. -P. Nozières, *IEEE Trans. Magnetics* **40**(4), 2625 (2004).

[42] N. Sakimura, T. Sugibayashi, T. Honda, H. Honjo, S. Saito, T. Suzuki, N. Ishiwata and S. Tahara, *IEEE J. Solid-State Circuits* **42**(4), 830 (2007).

[43] H. W. Schumacher, C. Chappert, R. C. Sousa, P. P. Freitas, J. Miltat and J. Ferré, *J. Appl. Phys.* **93**, 7290 (2003).

6   Spin-torque-transfer mode MRAM

6.1   Introduction

The best description of spin-torque transfer can be found in the patent issued to
John Slonczewski of IBM [1]:

*It is a fundamental fact that the macroscopic magnetization intensity of a magnet such as iron
arises from the cooperative mutual alignment of elementary magnetic moments carried by
electrons. An electron is little more than a mass particle carrying an electrostatic charge, which
spins at a constant rate, like a planet about its axis. The electric current of this spin induces
a surrounding magnetic field distribution resembling that which surrounds the Earth. Thus,
each electron is effectively a miniscule permanent magnet...*

*... The exchange interaction is that force, arising quantum-mechanically from electrostatic
interactions between spinning electrons, which causes this mutual alignment ... Not only does it
couple the bound spins of a ferromagnet to each other, but it also couples the spins of moving
electrons, such as those partaking in current flow, to these bound electrons.*

The subject of spin-torque transfer was not widely known until 1996. Due to its
enormous technology potential, both academic and industrial research activities
had been very active, and very rapid progresses have been made in recent years:
from the first experimental verification of spin-torque transfer in giant magnetoresistance (GMR) film, to the implementation of this mechanism to magnetic
tunneling junction devices. A large portion of this effort was directed towards
the development of practical magnetic RAM chips based on the spin-torque-
transfer mechanism. This chapter covers the physics of spin-torque transfer and
the switching properties of the magnetic tunnel junction, and also the technological aspects of spin-torque-transfer magnetic RAM technology.

6.2   Spin polarization of free electrons in ferromagnets

Electrons passing through a normal metal (NM), meaning non-ferromagnetic,
film are scattered in the film and lose their original spin orientation. The scattering
is independent of the spin polarization of the electron. Statistically, electrons
leaving a NM are made up of equal populations of the two spins. However, when
electrons enter a ferromagnetic (FM) layer from a NM layer or leave a FM layer
to go into a NM layer, their spins are affected by the ferromagnet. At the interface

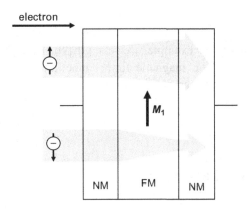

electron

M_1

NM   FM   NM

Figure 6.1. A ferromagnetic thin film polarizes itinerary conduction electrons and acts as a polarization filter.

of the FM and NM films, spin-dependent scattering takes place. Electrons with the same spin polarity as the FM layer (called majority-spin electrons) scatter less frequently at the interface than those with the opposite spin polarity (minority-spin electrons). This is the well known giant magnetoresistance (GMR) effect. As a result, a higher population of the electrons leaving the FM layer are polarized with majority spin.

In 1994, Y. Lassailly *et al.* showed that very thin Au–Co–Au film (Co ~ 2 nm) acts like a spin filter. The transmission coefficient of minority-spin electrons is about 0.7 times that of majority-spin electrons [2]. The polarization factor P is defined as $P = (N^+ - N^-)/(N^+ + N^-)$, where N^+ and N^- are the populations of electrons with majority spin and minority spin, respectively. For the Lassailly experiment, assuming that the population of the majority-spin electron is 1 and that of the minority-spin electron is 0.7, one obtains a polarization factor $P = (1 - 0.7)/(1 + 0.7) \sim 0.18$. Although this value is smaller than those in the recent literature, $P \sim 0.35$ for Co [3], the filter action is evident. Among ferromagnetic films, CoFeB is known to have the highest polarization factor, $P \sim 0.65$ [4], followed by Fe, $P \sim 0.40$. The spin-filter phenomenon is illustrated in Fig. 6.1.

The macroscopic parameter magnetization of a FM film is the net moment of the bounded, orbital electrons in the film. Such electrons do not contribute to the electrical conductivity, but do interact with the free electrons through the exchange of magnetic moment. Thus, the spin of the free electron is polarized by the polarity of the magnetization of a FM film. This process can be reciprocal. Namely, if a stream of polarized electrons is injected into a FM film, through the moment exchange, the polarization of the magnetization of the FM film may switch to the polarization of the free electrons. This concept was first described by Slonczewski [5] and separately by Berger [6]. The switching occurs when the polarized electron current density is greater than a threshold value, which was predicted to be in the order of mid-10^6–mid-10^7 A/cm^2 [6]. This phenomenon is called spin-torque-transfer switching, or spin-moment-transfer switching.

The verification of the spin-torque-transfer switching was first carried out in GMR film stacks in the early days. The switching-current density is in the range from mid-10^6 to mid-10^9 A/cm^2 (for example, see refs. [7]–[9]). Recently, studies of spin-torque-transfer switching on magnetic tunnel junctions show that the switching-current density is in the range of mid-10^6 A/cm^2.

For a comprehensive review of the physics of spin-torque transfer, see refs. [10] and [11].

6.3   Interaction between polarized free electrons and magnetization – macroscopic model

Like the coherent-rotation model (also called the Stoner–Wohlfarth model) in Chapters 2 and 3, the macroscopic model also treats the magnetization M_S of an entire film as one uniform magnet. While the coherent-rotation model does not describe transient behavior, the macroscopic model does. Thus, it can be used to analyze the dynamics of the behavior of the magnetization of a FM film. As mentioned in Chapter 2, the magnetization M_S is the volume density of the net moment of the uncompensated magnetic moment of the bounded electrons. One comprehensive work on the macroscopic model was published in 2000 [12], and readers are referred to that article for more details of the analysis.

Consequently, once the FM film is treated with one single macroscopic parameter, M, the analysis of the exchange interaction between the free electrons and the bounded electrons is greatly simplified. The quantitative result of this approach remains challenging, especially when the edge effect is dominant [13]. On the other hand, the microscopic model, which takes into account the interactions between layers of magnetic material and edge effects in three dimensions, provides a quantitatively more accurate prediction of the dynamic behavior, at the expense of very long computing time [14]. Nonetheless, the macroscopic model is adequate to introduce the concepts for tutorial purposes. Parts of the subsequent analyses are based on the macroscopic model to introduce the concept of spin torque and its effects on the switching properties of magnetic tunnel junctions.

As mentioned in Chapter 2, the dynamic behavior of M in a FM film can be described by the Landau–Lifshitz–Gilbert (LLG) equation, which can be re-written as follows:

$$\left(\frac{1}{\gamma}\right)\frac{dM}{dt} = (M \times H_{\text{eff}}) - \alpha \frac{M}{M} \times (M \times H_{\text{eff}}), \tag{6.1}$$

where γ is the gyro-magnetic ratio, $\gamma = 2\mu_B/\hbar$, α is the damping constant and H_{eff} is the effective field. The effective field can be the anisotropy H_K and/or the external field H that exists in the ferromagnet. The direction of $M \times H_{\text{eff}}$ is normal to the plane of M and H_{eff}. When α is zero, M precesses continuously around H_{eff} at a frequency $\omega = \gamma \cdot H$. Figure 6.2 illustrates the motion of spin precession. The

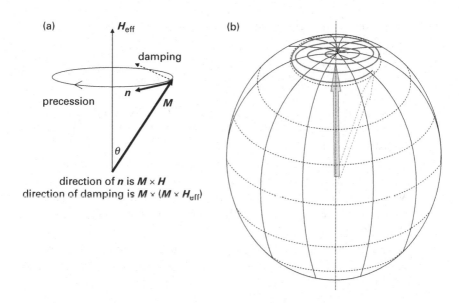

(a)

H_{eff}

damping

precession

n

M

θ

direction of n is $M \times H$
direction of damping is $M \times (M \times H_{eff})$

(b)

Figure 6.2. (a) Precession of spin angular momentum around a reference magnetic field H_{eff}. (b) The damping dissipates the energy of the precession moment, which spirals towards the reference field.

second term of the LLG equation is a vector, $M \times (M \times H_{eff})$ pointing from vector M toward field vector H_{eff}. When the damping constant α is greater than zero, the precession gradually dissipates its energy and the precession cone angle θ decreases. Eventually, M aligns to H_{eff}.

When a large flux of polarized electrons is injected into a ferromagnet, the dynamic motion of the magnetization given in Eq. (6.1) is modified: M not only experiences the usual anisotropy field and the external field in Eq. (6.1), it also experiences the torque from the passing-by polarized free electrons through the exchange interaction.

The electron flux of a current I is AJ/e, where A is the area of the FM film, e is the electron charge and J is the current density. Each free electron carries a quantized angular momentum of $\hbar/2$. The net moment available from the flux of free electrons, m_{el}, is $A(JP/e)(\hbar/2)$, since the electrons are partially polarized and $P < 1$. The passage of each spin-polarized electron rotates the spin momentum of the FM film by one quantum unit. In the FM layer, there are m_{fm} totally uncompensated local moments. Note that m can be referred back to the magnetization as the film volume V multiplied by the magnetization, i.e. $m_{fm} = VM_S = (At)M_S$, where t is the film thickness. The moment of the free electron m_{el} acts on the moment m_{fm} of the FM film through exchange. Thus, on average, each magnetization of the FM film shares a moment $m_{el}/m_{fm} = \hbar JP/2etM_S$ from the flux of free electrons.

The effect of the moment exchange on the dynamics of magnetization can be treated as an effective field, similar to the actual field in the ferromagnetic film.

Thus, one can include another current-density dependent term in the LLG equation as follows:

$$\left(\frac{1}{\gamma}\right)\frac{d\boldsymbol{M}}{dt} = (\boldsymbol{M} \times \boldsymbol{H}_{\text{eff}}) - \alpha \frac{\boldsymbol{M}}{M} \times (\boldsymbol{M} \times \boldsymbol{H}_{\text{eff}}) - \alpha_J \frac{\boldsymbol{M}}{M} \times (\boldsymbol{M} \times \boldsymbol{s}) \qquad (6.2a)$$

where s is the unit vector of the moment of the flux of polarized free electrons and $a_J = \frac{\hbar J P}{2etM_S}$. The direction of s is arbitrary. Without s, the third term in Eq. (6.2a) vanishes and M precesses around $\boldsymbol{H}_{\text{eff}}$.

When polarized electrons are injected into a FM film, $s > 0$ and the third term of Eq. (6.2a) is non-zero. Consider two special cases. The first case is that M (or $\boldsymbol{H}_{\text{eff}}$) and s are parallel. Then, $(\boldsymbol{M} \times \boldsymbol{H}_{\text{eff}})$ and $(\boldsymbol{M} \times \boldsymbol{s})$ have the same sign, the torque is in the same direction and, thus, the second and third terms both damp the precession of the moment.

The second case is that M (or $\boldsymbol{H}_{\text{eff}}$) and s are antiparallel, so the third term counteracts the damping. The second and third terms can be combined as follows:

$$\left(\frac{1}{\gamma}\right)\frac{d\boldsymbol{M}}{dt} = (\boldsymbol{M} \times \boldsymbol{H}_{\text{eff}}) - \alpha\left(1 - \frac{\alpha_J}{\alpha}\right)\frac{\boldsymbol{M}}{M} \times (\boldsymbol{M} \times \boldsymbol{H}_{\text{eff}}). \qquad (6.2b)$$

The effective damping is reduced by the current density of the polarized electrons.

Thus, at a certain flux density of the polarized free electrons, the effective damping becomes negative. Figure 6.3 illustrates the dynamic motion of the magnetization when a flux of polarized electrons is injected. At a moderate flux level, the magnetization adjusts its precession cone angle θ and eventually settles at an equilibrium angle; the precession is sustained and $1 - \frac{\alpha_J}{\alpha} = 0$. Such a dc-current-driven ferromagnetic resonance emits a radio frequency (RF) signal. The sustained precession and its RF emission was first predicted theoretically and observed by Berger [6]. At a higher flux level, $1 - \frac{\alpha_J}{\alpha} < 0$, the magnetization spirals away from the precession axis, the cone angle grows beyond $\pi/2$, and this results in magnetic reversal. The latter case is called the spin-torque-transfer switching.

6.4   Spin-torque transfer in a multilayer thin-film stack

Consider a five-layer alternating normal metal and ferromagnetic film stack (NM1–FM1–NM–FM2–NM2), as shown in Fig. 6.4. Suppose that there is an angle θ between the magnetizations M_1 and M_2 of layers FM1 and FM2, respectively. Free electrons, with equal populations of the two spins, enter the film stack from the left-hand side. When the electrons pass through the FM1 layer, FM1 acts as a spin filter. When leaving FM1, more free electrons are oriented to the direction of FM1. Since the NM layer is very thin, the polarized electrons maintain their spin polarization. When polarized electrons enter FM2, spin-torque transfer takes place, and M_2 experiences a torque in the direction of M_1. Thus, a flux of electrons from FM1 to FM2 can transfer the moment from M_1 to M_2 in such a film-stack structure.

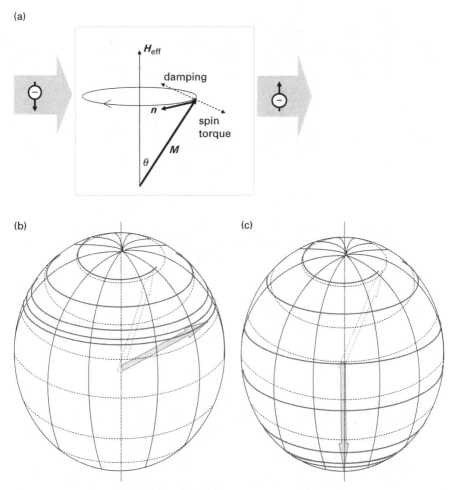

Figure 6.3. (a) Polarized free electrons exchange moments with the magnetization of opposite polarity, exerting a torque in a direction opposite to the direction of damping. (b) When the current is below a threshold value, the precession angle θ first increases and then reaches a steady angle. (c) When the current is greater than a threshold value, the precession angle θ continues to grow beyond $\pi/2$ and magnetic reversal takes place.

Similarly, reversing the direction of electron flow in Fig. 6.4 will reverse the process. Electrons are polarized by FM2, after which they enter FM1 and transfer their spin momentum to M_1.

Now consider the case of a GMR structure in which M_1 is fixed (cannot rotate) and M_2 is free to rotate. The GMR is initially in the antiparallel (AP) state, as shown in Fig. 6.5(a). Electrons coming from the left-hand side are filtered by FM1; they transfer their torque to FM2 and, if the current density is above a certain threshold, switch M_2 into the parallel (P) state.

On the other hand, the GMR film stack is initially in the P state, as shown in Fig. 6.5(b). Electrons coming from the right-hand side are filtered by M_2

electron

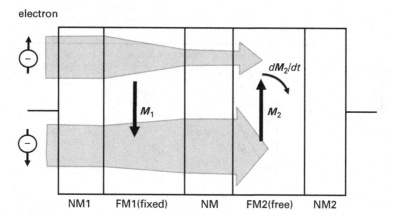

NM1   FM1(fixed)   NM   FM2(free)   NM2

Figure 6.4. A five-layer film stack NM1–FM1–NM–FM2–NM2, where FM = ferromagnetic material and NM = normal metal. (a) Free electrons enter from the left-hand side. They are polarized by FM1 and then enter FM2. The polarized electrons exchange momentum with the magnetization in FM2.

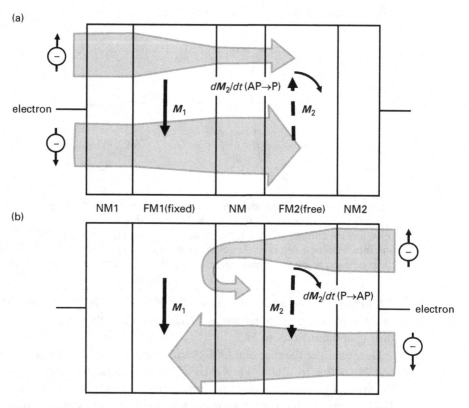

Figure 6.5. Current-induced switching in free ferromagnetic layer of a film stack. (a) Switching from antiparallel (AP) to parallel (P) state by the polarized electrons from the left. (b) Switching from P to AP state by the electron from the right-hand side and reflected from the NM–FM1 interface.

(thus with same polarization as M_2). The minority-polarized electron leaving FM2 is reflected back at the interface of NM and FM1. Both the incoming and the reflected minority-polarized electrons exchange with FM2. When the electron flux level is greater than a certain threshold, the free layer M_2 switches and the GMR film turns into an AP state.

6.5 Spin-transfer torque and switching threshold current density

Allowing for the partial polarization in the electron flux, the amplitude of the spin torque can be derived as follows [5]:

$$|\Gamma| = \left(\frac{Jg(\theta)}{e}\right)\hbar, \qquad (6.3)$$

where J is the current density, e is the electron charge, Jg/e means the effective flux of the polarized electrons from NM to FM2 and $\hbar$ is Planck's constant. Note that $g(\theta)$ is a function of the polarization factor P of the polarized electrons and the angle θ between the polarized electrons and M_2 of the FM2 layer; it is equal to

$$[-4 + (1+P)^3(3+\cos\theta)/4P^{3/2}]^{-1}.$$

The relationship is shown in Fig. 6.6 as a function of instantaneous angle θ; $g(\theta)$ vanishes at an angle $\theta = 0$ and π, for which the component of transferred spin oriented orthogonal to M_2 vanishes. Although the torque is zero when M_1 and M_2 are parallel or antiparallel, in practice the torque exists. The misalignment of the easy axis, as well as the thermal agitation at room temperature, cause M to deviate from perfect alignment. For small P, the transfer rate plotted in Fig. 6.6 turns into $\sim\sin\theta$.

At any instant in time, the torque from the free electron to M_2 pulls the magnetization away from the precession axis, and so changes its direction

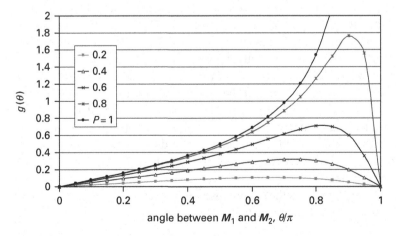

Figure 6.6. The instantaneous angle between the polarized electrons and M_2 vs. $g(\theta)$; P is the polarization factor, in the range 0.2–1.0. (After ref. [1].)

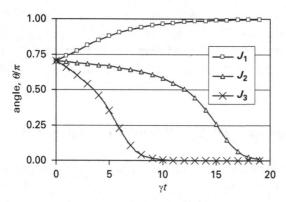

Figure 6.7. Normalized precession cone angle (θ/π) between two layers FM1 and FM2 vs. normalized transition time (γt). The solution of Eq. (6.4) with initial angle 0.7π is given, where $J_1 < J_2 < J_3$.

continually with time. The torque is proportional to the flux of the polarized free electrons entering FM2. As moment exchange takes place, the spin precession cone angle of FM2 increases until a new equilibrium cone angle is reached.

Consider a case with internal field $H = H_K$. The behavior of the precession cone angle is determined by the last two terms of Eq. (6.2b) and can be derived as follows [15]:

$$\frac{d\theta}{d(\gamma t)} = -\frac{1}{2}\alpha H_K \sin 2\theta - \left(\frac{\hbar J g(\theta)}{eM_2 t_2}\right) \sin \theta. \tag{6.4}$$

The precession cone angle θ responds more quickly to the density of the polarized current when the damping constant and magnetization are both small. Figure 6.7 shows the changes in the FM2 precession cone angle as a function of normalized time for three current densities, $J_1 < J_2 < J_3$, and the initial cone angle is 0.7π; H_K is in the direction of $\theta = \pi$. When the current density (J_1) is small, the spin torque is insufficient to overcome the torque from damping, the cone angle gradually settles at $\theta = \pi$, and M_2 aligns to H_K. When the current density (J_2, J_3) is larger than a certain threshold, the cone angle shrinks (actually M_2 moves away from H_K) and M_2 eventually aligns to the direction of the spin moment of the incoming polarized electrons. A larger current accelerates the switching process.

Such transient behavior of the magnetization was verified by Koch on a Co/Cu/ Co film stack [16]. The measurement result indicates that the time required to complete magnetic reversal is highly current-dependent. Figure 6.8 shows the transient behavior of the magnetization and the extracted transition time constant (this was referred to as decay time, τ_{DECAY}, in the paper). There are two transition-time regimes.

The first regime is the sub-nanosecond switching time τ at a current density over a critical value I_{C0}, in which

$$\tau^{-1} \approx \frac{P \cdot (\mu_B/e)}{m \cdot \ln(\pi/2\theta_0)}(I - I_{C0}), \quad \text{when } I > I_{C0}, \tag{6.5a}$$

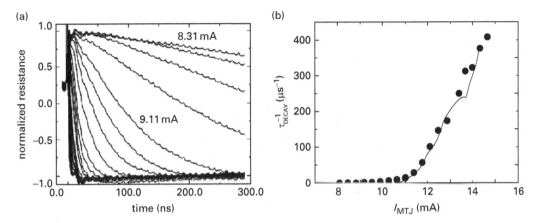

Figure 6.8. (a) Transient behavior of magnetoresistance of a Co/Cu/Co GMR film stack under various switching currents. (b) Extracted transition time (decay time) showing two distinct regimes. (After ref. [16], with permission from APS.)

where P is the polarization factor of the incoming electrons, θ_0 is the initial angle of deviation from the easy axis, $m = M_S V$, V is the volume of the free layer and

$$I_{C0} = (\alpha/P)(2e/\hbar)m(H + H_{\mathrm{K}} + 2\pi M_{\mathrm{S}}) \qquad (6.5b)$$

is the 0 K-temperature threshold current for spin-torque transfer (STT) magnetic reversal.

The second is a long-switch-time regime (>1 ns) in which the current is below I_{C0}:

$$\tau^{-1} \approx \tau_0^{-1} \exp[-\Delta_0(1 - I/I_{C0})], \text{ when } I < I_{C0}, \qquad (6.6)$$

where $\Delta_0 = E_{\mathrm{b}}/k_{\mathrm{B}}T$ is the normalized switching-energy barrier between the AP and the P state, which is equal to the anisotropy energy, K_{u}, when all other forces are absent.

6.6   Switching characteristics and threshold in magnetic tunnel junctions

The film stack structure of a magnetic tunnel junction is magnetically similar to the GMR structure. By replacing the middle NM layer with an insulator, the middle three layers of the structure in Fig. 6.4 become a magnetic tunnel junction, having FM1 as the pinned layer and FM2 as the free layer. Thus, the previous analysis can be applied to the MTJ. In this book, we will use "W1" to designate the transition of MTJ from the parallel (P) to the antiparallel (AP) state, and "W0" to denote the transition from the AP state to the P state. Figure 6.9 illustrates the I–V characteristics of MTJ when switching takes place.

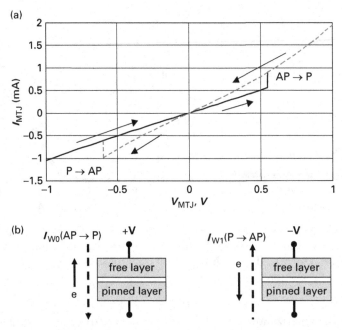

Figure 6.9. The I–V characteristics and convention of a magnetic tunnel junction. The device current changes abruptly when magnetic reversal takes place. The resistance of the P state is lower, and the current is larger.

6.6.1   Regimes of write pulse width

As discussed in Sections 6.4 and 6.5, Eqs. (6.5) and (6.6) show that the transition time of STT switching is current-dependent. The threshold current density of the spin-torque-transfer switching is current-pulse-width-dependent. Figure 6.10 shows the dependence of the write current as a function of the write pulse width. Again, there are two write pulse-width regimes, and their boundary in time is at a few nanoseconds.

In the very short pulse regime, $\sim$1 ns and shorter, the switching threshold increases rapidly as the pulse width shortens. In this pulse-width regime, the switching is achieved by the injection of a large polarized current that is sufficient to grow the precessional cone angle continuously till the magnetization flips, as described in Sections 6.4 and 6.5. This regime is called the precessional regime, and the empirical data fit well with the precessional model [12]. The switching-current dependence on the pulse width is given by

$$J_C = J_{C0}\left\{1 + \frac{\tau_{\mathrm{relax}}}{t_{\mathrm{PW}}}\ln\left(\frac{\pi/2}{\theta_0}\right)\right\}, \tag{6.7}$$

where t_{PW} is the pulse width, and τ_{relax} and θ_0 respectively represent the relaxation time and the root square average of the initial angle of the free-layer magnetization, which is determined by thermal fluctuation. Data in the $<$10 ns regime can be fitted well to this model [17, 18].

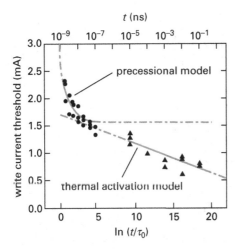

Figure 6.10. Write-current threshold as a function of write-current pulse width. The slope gives the thermal barrier. For $\tau_0 = 10^{-9}$ s, $I_{C0} = 1.7$ mA and $E_b = 32\, k_B T$ are obtained in the long pulse regime; $\tau_{\text{relax}} = 3 \times 10^{-10}$ s. (After ref. [18].)

In the long pulse width regime, the switching may take place at levels of polarized current that barely sustain the precession of the magnetization, but are insufficient to grow the precession angle beyond $\pi/2$. Nonetheless, when the magnetization precesses with a finite cone angle, its switching-energy barrier is lowered, and thus the switching probability increases. When the write pulse is long enough, the accumulated switching probability approaches unity and the switching takes place. This regime is also called the thermal switching regime, in which the switching current threshold decreases with the logarithm of the pulse width t_{PW} as

$$J_C = J_{C0}\left\{1 - \left(\frac{k_B T}{E_b}\right)\ln\left(\frac{t_{PW}}{\tau_0}\right)\right\}, \tag{6.8}$$

where J_C is the threshold current, J_{C0} is the switching threshold current at 0 K, E_b is the switching-energy barrier between the two (AP and P) states, t_{PW} is the pulse width and τ_0 is the inverse of the spin-reversal attempts in units of time (seconds). As shown in Fig. 6.10, the slope of J_C vs. ln (t_{PW}) is $k_B T/E_b$.

6.6.2 Switching probability in the thermal regime

In the thermal regime, $J < J_{C0}$, the switching probability is finite over a range of write current values as shown in Fig. 6.11. Since the probability of magnetization reversal is dependent on the switching-energy barrier between the two states, the switching probability is expressed as the following [16, 19]:

$$P_{SW} = 1 - \exp\left\{-\frac{t_{PW}}{\tau_0}\exp\left[-\frac{E_b}{k_B T}\left(1 - \frac{I}{I_{C0}}\right)\right]\right\}, \tag{6.9}$$

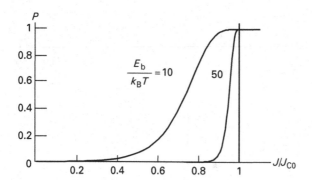

Figure 6.11. The probability of write success vs. normalized switching current density as a function of energy barrier; based on Eq. (6.9).

where t_{PW} is the pulse width and I_{C0} is the threshold write current. The probability increases sharply when the write current approaches the threshold I_{C0}. The probability increases with write pulse width. Note that the switching-energy barrier also plays a role in the sharpness of the transition. The larger the switching-energy barrier, the more abrupt is the transition from low to high switching probability. Of course, this is true only when the temperature of the MTJ remains constant.

6.6.2.1   Joule heating of the tunnel barrier

When the large current flows in the MTJ during the spin-torque-transfer write process, it is expected that the MTJ temperature rises. We can set a homework problem on the temperature rise in such a write operation.

Homework

Q6.1 Estimate the temperature increase and the magnetic field in a MTJ with a 1 nm thick tunnel barrier and a circular area with a diameter of 100 nm. Assume that the overall thermal resistivity of the MTJ is $\kappa = 1.5\,\text{m} \cdot \text{K/W}$ (where m = meter, K = kelvin). The applied voltage is 1.0 V and $RA = 7.85\,\Omega\,\mu\text{m}^2$ at bias = 1 V, where RA is the product of the MTJ resistance and area.

A6.1 The temperature rise may be calculated as follows:

MTJ area $A = \pi(0.1\,\mu\text{m}/2)^2 = 7.85 \times 10^{-3}\,\mu\text{m}^2 = 7.85 \times 10^{-15}\,\text{m}^2$;

MTJ resistance $= 7.85\,\Omega\,\mu\text{m}^2/7.85 \times 10^{-3}\,\mu\text{m}^2 = 1\,\text{k}\Omega$;

MTJ current $= 1\,\text{V}/1\,\text{k}\Omega = 1\,\text{mA}$.

The power into the MTJ is given by $P = 1\,\text{mW}$ across the tunnel barrier; the rest of the MTJs are metallic and of low resistance, thus it generates little power.

The power consumed by the MTJ is $P = 1\,\text{V} \cdot 1\,\text{mA} = 1\,\text{mW}$. Since the MTJ electrodes are metallic, and are therefore of low ohmic resistance and

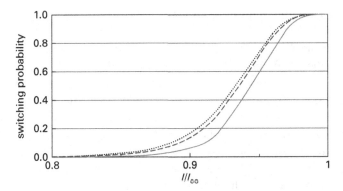

Figure 6.Q1. Switching probability vs. write current for $E_b/k_BT = 40$ (dotted line) and $E_b/k_BT = 50$ (solid line) at constant temperature. The dashed line $E_b/k_BT = 50$ includes Joule heating. Pulse width $= 10$ ns, and we assume that $\Delta T = 0.2\, I/I_{C0}$.

low thermal resistance, heat is generated at the tunnel barrier and is then conducted away via electrodes on both sides of the MTJ. The distance from the middle of the tunnel barrier to the electrode is given by $d = 0.5t_{tb} = 0.5$ nm, where t_{tb} is the thickness of the tunnel barrier. Thus the thermal resistance from the middle of the tunnel barrier to the two electrodes is given by

$$\varsigma = [(d/2)/A] \cdot \kappa = [0.25 \times 10^{-9}\,\text{m}/(7.85 \times 10^{-15}\,\text{m}^2)]$$
$$\times 1.5\,\text{m} \cdot \text{K/W} = 4.8 \times 10^4 (\text{K/W}).$$

Thus, the temperature rise is given by

$$\Delta T = W\varsigma = (1\,\text{mW}) \times 4.8 \times (10^4\,\text{K/W}) = 48\,\text{K}.$$

From Ampère's Law, the magnetic field at the edge of the MTJ is $H = I/2\pi r = 1\,\text{mA}/(2\pi \times 50 \times 10^{-9}\,\text{m}) = 3.184\,\text{kA/m} = 39.8\,\text{Oe}$ in the circular direction. The magnetic field is smaller in the middle of the MTJ.

The Joule heating of the tunnel barrier raises the temperature of the entire MTJ, including the free layer. The switching current across the MTJ generates a field. Equation (6.9) has been used to extract $E_b/k_BT = 10$. Figure 6.Q1 shows that the method is affected by Joule heating, and gives a lower than actual value for E_b.

Similarly, when the heating of the MTJ during the write operation is not negligible, the switching-energy barrier, E_b, extracted from Eq. (6.8) is lower than the actual value. The switching current at different write pulse widths heats up the MTJ to different temperatures. The temperature at each point of Fig. 6.Q1 is different: in the shorter pulse regime, more current is applied to switch, and the temperature of the MTJ is higher. Unless an accurate temperature is taken into account, E_b cannot be extracted accurately. It has been shown that the switching-energy barrier extracted from this method is smaller than that using field switching. We will introduce another E_b extraction method later that is taken at constant temperature.

6.6.2.2 Write error rate and read disturb rate

The write error rate can be experimentally measured by repeatedly switching the MTJ between the AP and P states for a particular pulse width. At the time of writing, a unified definition of switching threshold has not been determined. One finds from the literature that it is becoming increasingly common to call the write current the "threshold current" at the 50% write success rate. Although previous switching threshold analyses are based on current density, in practice it is more convenient to measure switching threshold voltage, similar to the threshold voltage of a transistor. From here on, this book will quote threshold in both current and voltage.

The write error rate and the read disturb rate are important parameters for MRAM design. They are used to project the write current and read current design point and product fail margin. These points will be discussed in the later part of this chapter.

6.6.2.3 Write error rate

The write voltage is proportional to the product of the MTJ resistance and area, RA, since $V_W = I_W R = (J_W A) R = J_W (RA)$, where V_W and I_W are write voltage and write current, respectively. Lowering the RA value of the MTJ will lower the write voltage while keeping the write current the same. The write bit error rate (WER) is given by

$$\text{WER} = 1 - P_{SW}. \tag{6.10}$$

Figure 6.12 shows a comparison of the WER of two kinds of MTJs with different values of RA. The correlation is clear. Since $V_W = RA \cdot J_W$, it is expected that the write voltage of MTJs with higher values of RA is greater.

The dependence of the write error rate on write pulse width is illustrated in Fig. 6.13. As one increases the switching current to raise the switching probability, the temperature of the MTJ gradually rises. So, the actual write error rate decreases slower than at a constant temperature. The write error rate of a long write pulse in the range 50–100 ns can be reasonably fitted with a *complement error function*,

$$\text{WER} = 1 - P_{SW} = erfc(V_w),$$

where V_W is the write voltage. The error function is the integration of a Normal distribution function from the center to the edge, $erfc(z) = \int_z^\infty \exp(-x^2)dx$. In this case,

$$x = \frac{V_W - V_{W50}}{\sqrt{2}\sigma_{\text{temporal}}},$$

where V_{W50} is the switching threshold voltage and σ_{temporal} is the temporal standard deviation of the write error rate.

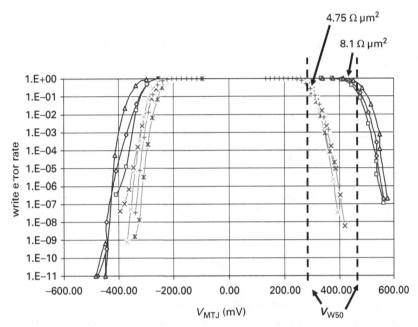

Figure 6.12. The write error rate (WER) of MTJs with $RA = 8.1$ and $4.75 \ \Omega \ \mu m^2$ (three samples given for each). The pulse width is 100 ns, and V_{W50} denotes the threshold voltage. (Courtesy of MagIC Technologies, Inc.)

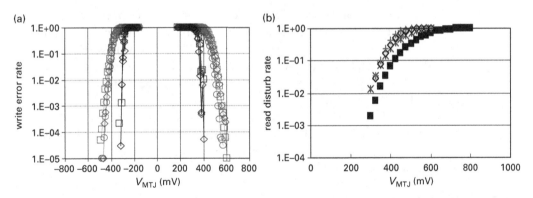

Figure 6.13. (a) Write error rate vs. write voltage of three MTJs for two write pulse widths, 100 ns and 9500 ns. The slope of the write error rate is the temporal standard deviation $\sigma_{temporal}$ of the switching, which is pulse-width dependent. (b) Read disturb rate ($= 1 - $ WER). (Courtesy of MagIC Technologies, Inc.)

When writing with very long pulse widths ($t_{PW} > 300 \ ns$), the dependence of the error rate on the write voltage, V_W, deviates from the Normal distribution and can be approximated by the integral Weibull distribution. It takes the form

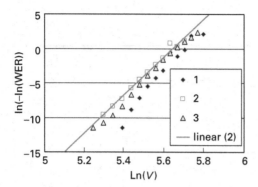

Figure 6.14. Weibull plot of write error rate in Fig. 6.12 for PW = 9500 ns; three samples (1, 2, 3). The solid line is a linear fit of sample 2. (Courtesy of MagIC Technologies, Inc.)

$$WER = \exp\left[-\left(\frac{V_{MTJ}}{V_0}\right)^{\beta}\right],$$

which is a straight line in a Weibull plot (see Fig 6.14 and Appendix E).

A wider write pulse gives rise to a steeper drop in the write error rate, or the σ_{temporal} is smaller. On the other hand, for shorter pulses, the drop is slower than the Normal function and the write error rate decreases more slowly than those for long pulses. Error rate measurements are required to make a projection of the switching window size for chip level design. It is possible that more transition mechanisms are involved.

Nonetheless, for practical device applications, the write error rate should reach below 10^{-9}. For a Normal distribution, this means that the write voltage must be large enough to cover $V_W = V_{W50} + 6\sigma_{\text{temporal}}$. The write voltage V_W is greater than the write threshold V_{W50}. As shown in Fig 6.12, for a 100 ns write pulse width, the V_W between the write threshold, V_{W50} (defined as the write error rate = 0.5) and WER = 10^{-9} is ~140 mV.

When an array of MTJs is considered, the spatial spread, σ_{spatial}, of the WER from MTJ to MTJ should also be considered. Thus, the total spread is $\sigma_{\text{total}} = (\sigma_{\text{temporal}}^2 + \sigma_{\text{spatial}}^2)^{1/2}$. From σ_{total}, one can project the write error rate of a chip at a given V_W.

6.6.2.4   Read disturb rate

To read a cell, a read voltage of the order of 0.1 V is applied to the MTJ to sense its resistance state. Although the read current is much smaller than the write threshold, there is a very low, but finite, probability that the MTJ may switch. Depending on the state of the MTJ, there is a 50% chance that the read current flows in the direction of lowering the switching-energy barrier, and this increases the probability of switching. If the unwanted switching occurs during read, it is called a read disturb. Using the same analysis as for the write bit error rate, the read disturb rate is also current dependent and is equal to P_{SW}.

Homework

Q6.2 For a MTJ with thermal factor $E_b/k_BT = 40$, what would be the difference in write current at error rates 0.5 and 10^{-9} if the write current pulse is (i) 100 ns long and (ii) 1 µs long? Assume that the temperature of the MTJ is constant.

A6.2 Start from Eq. (6.9). The thermal attempt switching time constant τ_0 is 1 ns. Note that the write error rate $(WER) = (1 - P_{SW})$. One rearranges Eq. (6.9) to obtain

$$I(WER) = I_{C0}\left\{1 + \frac{k_BT}{E_b}\ln\left[-\frac{\tau_0}{t_{PW}}\ln(WER)\right]\right\}.$$

Rearranging the $I(WER)$ terms, one obtains the difference in current at two different error rates as follows:

$$[I(WER1) - I(WER2)]/I_{C0}$$

$$= \frac{k_BT}{E_b}\left\{\ln\left[\frac{\tau_0}{t_{PW}}\ln(WER2)\right] - \ln\left[\frac{\tau_0}{t_{PW}}\ln(WER1)\right]\right\}$$

$$= \frac{k_BT}{E_b}\left\{\ln\left[\frac{\tau_0}{t_{PW}}\ln(WER2) \div \frac{\tau_0}{t_{PW}}\ln(WER1)\right]\right\}$$

$$= \frac{k_BT}{E_b}\{\ln[\ln(WER2)/\ln(WER1)]\},$$

which is independent of t_{PW} but is inversely dependent on the thermal barrier. Numerically, for a pulse width of 100 ns,

$$I(WER = 0.5) = I_{C0}\left\{1 + 0.025\ln\left[-\frac{1}{100}\ln(1 - 0.5)\right]\right\} = 0.876I_{C0}$$

and

$$I(WER = 10^{-9}) = I_{C0}\left\{1 + 0.025\ln\left[-\frac{1}{100}\ln(1 - 10^{-9})\right]\right\} = 0.961I_{C0},$$

thus the difference is $(0.961 - 0.876)\,I_{C0} = 0.085I_{C0}$.
For a pulse width of 1 µs,

$$I(WER = 0.5) = I_{C0}\left\{1 + 0.025\ln\left[-\frac{1}{1000}\ln(1 - 0.5)\right]\right\} = 0.818I_{C0}$$

$$I(WER = 10^{-9}) = I_{C0}\left\{1 + 0.025\ln\left[-\frac{1}{1000}\ln(1 - 10^{-9})\right]\right\} = 0.903I_{C0}.$$

The difference is again $0.085I_{C0}$.

The temporal standard deviation of the WER is write-pulse-width-dependent. This simple analysis does not include this term in the I/I_{C0} term.

Q6.3 For a MTJ with thermal factor $E_b/k_BT = 40$, what would be the difference in write current at an error rate of 10^{-9} and a read disturb rate of 10^{-9} if the write current pulse is 100 ns long and the read current pulse is 10 ns long?

A6.3 Following Q6.2, the write current is $0.961I_{C0}$. The read disturb rate is P_{SW}, thus

$$I(\text{WER} = 1 - 10^{-9}) = I_{C0}\left\{1 + (1/40) \cdot \ln\left[-\frac{1}{10}\ln(1 - 10^{-9})\right]\right\} = 0.425I_{C0}.$$

The difference between the write and read currents is given by

$$I_W - I_R = (0.961 - 0.425)I_{C0} = 0.536I_{C0}.$$

Q6.4 Same as Q6.2, but with $E_b/k_BT = 60$.

A6.4 The write current is $0.974I_{C0}$ and the read current is $0.616I_{C0}$, thus $I_W - I_R = 0.358I_{C0}$. Comparing answers from Q6.2 and Q6.3, it is clear that the gap between the read voltage and the write voltage is smaller for MTJs with larger switching barriers.

Q6.5 Prove that the relation between the write error rate of two write pulse widths as a function of I/I_{C0} is a power law.

A6.5 From Eq. (6.10),

$$\text{WER}(PW2) = \exp\left\{-\frac{t_{PW2}}{\tau_0}\exp\left[-\frac{E_b}{k_BT}\left(1 - \frac{I}{I_{C0}}\right)\right]\right\}.$$

Since $\exp(a \cdot b) = [\exp(b)]^a$,

$$\text{WER}(PW1) = \exp\left\{-\left(\frac{t_{PW1}}{t_{PW2}}\right)\frac{t_{PW2}}{\tau_0}\exp\left[-\frac{E_b}{k_BT}\left(1 - \frac{I}{I_{C0}}\right)\right]\right\}$$

$$= \exp\left\{-\frac{t_{PW2}}{\tau_0}\exp\left[-\frac{E_b}{k_BT}\left(1 - \frac{I}{I_{C0}}\right)\right]\right\}^{(PW1/PW2)}$$

$$= \text{WER}(PW2)^{(t_{PW1}/t_{PW2})}.$$

The WER falls much more slowly when the pulse width is short. In other words, the σ_{temporal} of the short write pulse is larger (see Fig. 6.13(a)).

6.6.3 Spin-torque-transfer switching under a magnetic field

In the thermal switching regime, the external magnetic field in the direction of the easy axis assists (or retards) the STT switching threshold [20, 21]. The field alters the switching-energy barrier in the same manner as in the field MRAM in the single-domain analysis. Combining the field term and the current term, one obtains the switching-energy barrier as follows:

$$\Delta = \Delta_0(1 - H/H_{C0})^2(1 - I/I_{C0}), \tag{6.11}$$

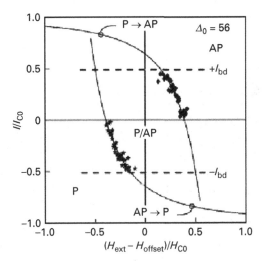

Figure 6.15. Phase diagram of switching field and write current: H_{ext} is the external field; H_{offset} is the offset of the R-H loop; H_{C0} is the critical (switching) field without current injection; I_{C0} is the critical STT switching current; I_{bd} is the breakdown current; and the switching-energy barrier under zero field and current is 56 $k_B T$. The lines depict the solution of Eq. (6.9) with the switching-energy barrier replaced by Eq. (6.11). (After ref. [20], with permission from AIP.)

where $\Delta_0 = E_b/k_B T$. Figure 6.15 shows the measured switching current in the presence of the easy-axis magnetic field. When the MTJ is injected with a write current to switch it from the P to the AP state, an external field in the AP direction lowers the energy barrier: $\Delta = \Delta_0 (1 - H/H_{C0})^2$. As a result, the switching current is reduced. On the other hand, when a field in the P direction raises the energy barrier, $\Delta = \Delta_0 (1 + H/H_{C0})^2$, the switching current increases.

6.6.4 Magnetic back-hopping

When a MTJ receives a polarized current in the direction that attempts to switch it to the same state (e.g., from an AP to an AP state), the MTJ should not switch. However, it has been observed experimentally that the MTJ may switch to the opposite state (in this case, to the P state). Such a phenomenon is called *back-hopping*. It happens frequently at currents greater than the write threshold, at a finite probability, mostly in samples with low anisotropy [22]. Figure 6.16 shows the back-hopping events at write voltages higher than the switching threshold. The state of the MTJ is indicated by the remnant current measured at $V_{MTJ} = 0.1$ V. A positive pulse writes the MTJ into the P state, and vice versa. The MTJ switches at 0.8~1 V, and the back-hopping occurs at a voltage greater than the switching voltage.

The back-hopping event occurs more often when the write pulse is short, around 10 ns, and the frequency of event occurrence decreases as the write pulse

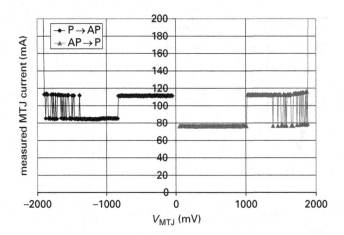

Figure 6.16. Remnant MTJ states vs. amplitude of write voltage pulse, showing back-hopping events occurring at amplitudes greater than the first switching voltage $\sim 0.8 - 1$ V. The MTJ breakdown voltage is ~ 1.8 V. (Courtesy of MagIC Technologies, Inc.)

width increases. The back-hopping can be minimized by increasing the anisotropy (or coercivity) of the MTJ [22]. Based on Eqs. (6.9) and (6.11),

$$P_{SW} = 1 - \exp\left\{ -\frac{t_{PW}}{\tau_0} \exp\left[-\Delta_0\left(1 - \frac{H}{H_{C0}}\right)^2\left(1 - \frac{I}{I_{C0}}\right)\right]\right\}.$$

When $|x| \ll 1, \exp(x) \approx 1 + x$, thus the switching probability

$$P_{SW} \approx \frac{t_{PW}}{\tau_0} \exp\left[-\Delta_0\left(1 - \frac{H}{H_{C0}}\right)^2\left(1 - \frac{I}{I_{C0}}\right)\right]$$

depends on the "energy barrier," which is modified by the applied current and field. The probability of forward-hopping (= switching) is given by

$$P_{SW-F} = P_0 \cdot \exp\left[-\Delta_0\left(1 - \frac{H}{H_{C0}}\right)^2\left(1 - \frac{I}{I_{C0}}\right)\right], \tag{6.12}$$

where the pulse-width term is replaced by P_0, and the probability of back-hopping is given by

$$P_{SW-B} = P_0 \cdot \exp\left[-\Delta_0\left(1 + \frac{H}{H_{C0}}\right)^2\left(1 + \frac{I}{I_{C0}}\right)\right]. \tag{6.13}$$

The ratio of probability for forward-hopping (= switching) and back-hopping is given by

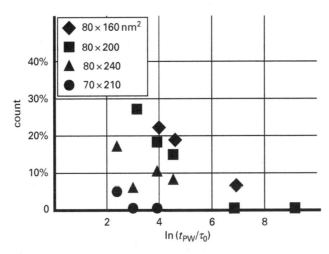

Figure 6.17. Occurrence frequency of magnetic back-hopping event in MTJ vs. write pulse width of MTJs with different length/width aspect ratio; $\tau_0 = 1$ ns. (Courtesy of MagIC Technologies, Inc.)

$$\frac{P_{\text{SW-B}}}{P_0} = \left(\frac{P_{\text{SW-F}}}{P_0}\right) \left(\frac{I_{C0} + I}{I_{C0} - I}\right) \left(\frac{H_{C0} + H}{H_{C0} - H}\right)^2. \tag{6.14}$$

Thus, $P_{\text{SW-B}} \ll P_{\text{SW-F}}$, but the probability is greater than zero. Increasing Δ_0 lowers the back-hopping probability. Figure 6.17 illustrates the occurrence frequency of the back-hopping event as a function of shapes. Back-hopping frequency is smaller in MTJs with larger length to width ratio, which exhibits higher coercivity and thus a higher switching-energy barrier.

6.7   Reliability of tunnel barriers in MTJs

Similar to the gate oxide (SiO_x) of Si field-effect transistors, the reliability of the magnetic tunnel barrier must be studied to ensure the long-term reliability of magnetic tunnel junctions as memory devices. The methods used to evaluate the reliability of tunnel barriers are time-dependent dielectric breakdown (TDDB) and resistance shifts [23–25]. For field MRAM, the *RA* value of the MTJ is higher, of the order of 1 kΩ μm^2, and the read current through the MTJ is small. It has been proven that reliable MTJs for field MRAM can be manufactured.

For the STT MRAM, the *RA* value of the MTJ is of the order 10 Ω μm^2, and the write current density is two orders of magnitude larger than the read current of the MTJs in a field MRAM. Thus, the tunnel barrier of the STT MTJ is stressed at a much higher level, and requires much higher integrity. Figure 6.18 shows commonly observed junction breakdown characteristics in low-*RA* samples. The distribution of breakdown shows the bimodal distribution of two overlapping

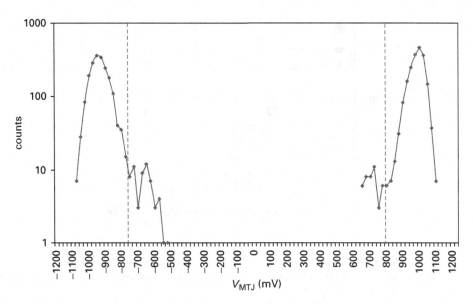

Figure 6.18. Breakdown voltage distribution of $8\,\Omega\,\mu m^2$ MTJs made of a MgO tunnel barrier. Note that, in addition to the main peak, there is a second peak with lower breakdown voltage. (Courtesy of MagIC Technologies, Inc.)

Normal distributions. The *I–V* curves of the higher breakdown group exhibit abrupt breakdown characteristics; this is called an intrinsic breakdown. The *I–V* curves of the lower group usually exhibit soft breakdown characteristics. The population of the lower breakdown group is heavily dependent on the tunnel barrier formation process. Thus, it is usually considered as comprising the extrinsic properties of the tunnel barrier, which are related to the defects in the tunnel barrier. The percentage of the soft breakdown increases as the *RA* value of the MTJ decreases.

The tunnel barriers that exhibit soft breakdown tend to be less reliable. After voltage stress over time, the tunnel current increases gradually, even if breakdown does not take place. Figure 6.19 shows tunnel current I_p vs. breakdown voltage of two groups of the same MTJs: one group was not stressed and the other was. The tunnel barriers that exhibit high breakdown voltage hardly change the distribution of tunnel current I_p after stress. Those in the middle range of breakdown voltage distribution show resistance drift after stress, and those in the lower range of the breakdown distribution become short and are off the chart.

6.8   SPICE model of MTJs and memory cells

The equivalent circuit of the MTJ is a binary resistor in parallel with a capacitor [26]. The tunneling current as a function of bias has been derived, as a polynomial of voltage in odd power, from a simple square potential barrier [27]:

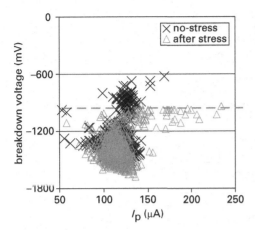

Figure 6.19. Soft breakdown and the degradation of the tunnel junction after stress. Voltage stress is conducted by a 600 mV, 100 ns pulse over 10^6 cycles. The MTJ tunnel current I_p is measured at $V_{MTJ} = 100$ mV prior to the breakdown measurements for both groups. (Courtesy of MagIC Technologies, Inc.)

$$I = a_1 V + a_3 V^3 + \cdots. \tag{6.15}$$

Thus, the direction of the current follows the polarity of the bias voltage, and the tunnel resistance is given by

$$R = \frac{V}{I} = 1/(a_1 + a_3 V^2 + \cdots). \tag{6.16}$$

The tunnel resistance decreases as the bias voltage increases. Depending on the details of the junction formation process, the resistance of the AP state drops more quickly as the voltage increases. For tunnel junctions with different ferromagnetic materials on the two sides of the tunnel barrier, the current is bias-polarity-dependent, and so is the tunnel resistance. At low bias, the tunnel current of a tunnel barrier is inversely proportional to $\ln(d)$, where d is the thickness of the tunneling barrier. One important parameter is V_{50}, the voltage at which the TMR ratio drops to 50% of the value at $V_{MTJ} \sim 0$.

The capacitance of the MTJ can be different from the geometrical dielectric capacitance, $C_d = \varepsilon \varepsilon_0 A/d$, where ε is the dielectric constant ($\varepsilon \sim 8.8$), ε_0 is the permittivity of a vacuum (8.86×10^{-14} farad/cm), A is the area and d is the thickness of the tunnel barrier. For a MTJ with Al_2O_3 barriers, the capacitance is much smaller than the geometric dielectric capacitance [28]; on the other hand, for MTJs with a MgO tunnel barrier, the capacitance is much larger than the geometric dielectric capacitance [29].

Temperature-dependent measurements of a MgO MTJ junction impedance has shown that the capacitance of the MTJ switches from high to low when the MTJ switches from a P to an AP state, opposite to the resistance change of the tunnel barrier. Additionally, for a P state, the capacitance varies with temperature,

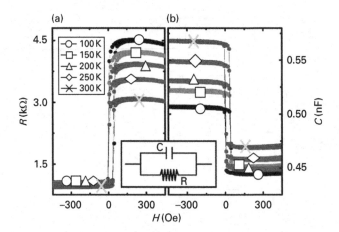

Figure 6.20. Junction resistance (a) and junction capacitance (b) of MgO (3 nm) MTJ at various temperatures. Note that both resistance and capacitance switch under an external magnetic field. (After ref. [29], with permission from AIP.)

although the resistance remains approximately constant. Figure 6.20 shows the capacitance and resistance of a MgO tunnel barrier.

The change in the capacitance is explained by screening due to charge and spin accumulation at the interfaces between the ferromagnetic electrode and the dielectric layer. The screening length can be positive or negative due to the oscillatory nature of the exchange interaction in addition to the Coulomb interaction, resulting in positive and negative interface capacitance C_i. The equivalent capacitance of the MTJ is therefore two capacitors in series: one is the geometrical capacitor C_d and the other is a state-dependent insulator–electrode interface capacitor C_i. A small difference in the screening length between the parallel and antiparallel orientations of the magnetization of the electrodes indicates the presence of charge and magnetization screening at the interfaces. Fig. 6.21 shows the equivalent circuit of the MTJ.

6.9   Memory cell operation

Figure 6.22 illustrates the circuit schematics of a 2×2 dual-bit-line STT MRAM cell array. Each cell is made up of a MTJ in series with a gating transistor, thus it is a 1T-1MTJ cell. The two ends of each cell are connected to a pair of bit lines, BLT (true) and BLC (complement). The gate of the transistor is connected to a word line. When the cell is selected, the word line is activated ($V_{WL} = V_{dd}$) and the transistor is turned on. Meanwhile, the corresponding bit-line pair is activated for read and write operation.

The read operation is the same as the field MRAM, in which a small sense voltage is applied to the cell and the resistance of the MTJ is sensed to determine the data stored in the cell. For a STT MRAM cell, the sense voltage across the

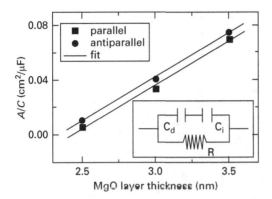

Figure 6.21. The inverse capacitance for P- and AP-state MgO MTJs with different MgO layer thickness. The inset shows the modified equivalent capacitance of the MgO layer, including geometric capacitance C_d and an interface capacitance C_i of the MgO tunnel barrier. In this case, C_i is negative. (After ref. [29], with permission from AIP.)

MTJ must be small enough such that the current in the cell does not disturb the content of the cell. A typical sense voltage is ∼0.1 V.

To write "0" into a cell, a positive write voltage, V_{CELL}, is applied to the BLT, the word line of the cell is activated ($V_{WL} = V_{dd}$), and the BLC is grounded. A current I_{W0} flows from the BLT via the MTJ and the transistor to the BLC. Electrons flow in the opposite direction to the current, in this case from the pinned layer to the free layer. As a result of spin-torque transfer, the free layer switches to the same polarization of the pinned layer, thus the MTJ is written into the P state. Conversely, to write "1" into a cell, the BLT is grounded, the BLC is raised to V_{CELL} and a write current I_{W1} flows in the reverse direction.

In the STT MRAM cell, the gating transistor must be able to provide a current large enough to switch the MTJ, which is close to one order of magnitude larger than the read current. The transistor current is proportional to the gate width. For a typical 90 nm NMOS transistor, the saturate drain current (maximum) $I_{ds,sat}$ is of the order of 800 μA/μm gate width. However, only a fraction of this current is available for write since the V_{ds} of the transistor is smaller than V_{dd}. Thus, the write current becomes the determining factor of the cell size. The smallest gate width of manufacturable transistors is in the 120 nm range, thus the maximum available write current for the smallest transistor is only about 95 μA. If the write current is greater than this value, the transistor size must be increased, and so must the cell size.

During the write "1" operation, the MTJ acts as a source follower to the transistor. The gate-to-source bias V_{gs} of the transistor is $V_{WL} - V_{MTJ}$, thus the gating transistor conducts less current than when $V_{gs} = V_{WL}$ for the write "0" operation. To compensate for the loss of V_{gs}, the gate voltage is frequently raised

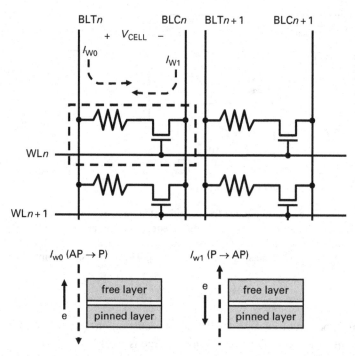

Figure 6.22. Dual bit line STT MRAM cells and the direction of write current.

above V_{dd}. This design technique is called "bootstrapping," and the circuit is called a "bootstrap circuit." Thus, the cell can provide more write current without increasing the size of the transistor. The downside of such a design technique is that the reliability of the transistor is compromised, and the transistor gate integrity degrades over time at a faster pace. The acceptable amount of boost in gate voltage is determined by the amount of degradation of the transistor at the end of the product life, which is determined by the kind of application.

6.9.1   *I–V* characteristics of STT memory cell during write

During the write operation, the transistor is driven into saturation, and its drain current characteristics are non-linear. Given a cell voltage V_{CELL}, the cell current can be obtained graphically using a "load line" method. The MTJ is the load resistor to the transistor.

Figure 6.23 illustrates the cell operation *I–V* characteristics when the cell is written with a constant voltage between BLT and BLC. The cell equivalent circuit is shown in Fig. 6.23(a). The MTJ and the transistor share the same cell current.

The $R_{MTJ} - V_{MTJ}$ trajectory is shown in Fig. 6.23(b). In this case, $+V_{MTJ}$ is applied first to write the cell from the AP to the P state (steps 1–3); then $-V_{MTJ}$

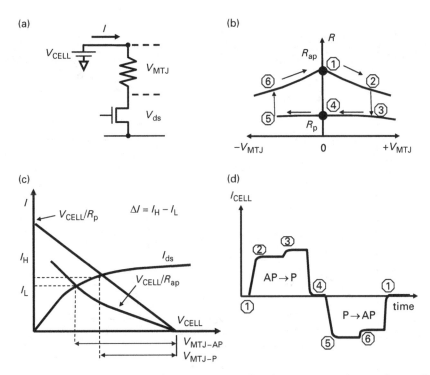

Figure 6.23. Write operation of STT MRAM cell under constant-voltage write condition: (a) equivalent circuit schematics; (b) MTJ resistance switching trajectory; (c) MTJ load lines for determining the cell currents; (d) current waveform when writing AP-to-P and P-to-AP.

is applied to write the MTJ from a P to an AP state (steps 4–6). Prior to the turn on of the transistor (step 1), there is no voltage across the MTJ, i.e. $V_{MTJ} = 0$ and $I_{CELL} = 0$. All the cell voltage drops across the drain and source of the transistor, and $V_{CELL} = V_{ds}$. When the cell is selected, its transistor is turned on, and current flows through the MTJ and a voltage exists across the MTJ (steps 2, 3).

When the transistor is on, V_{MTJ} and I_{MTJ} can be determined using the load line method, as shown in Fig. 6.23(c), since $V_{CELL} = V_{ds} + V_{MTJ}$ and the current is the same. Note that I_{ds} is the transistor current and the two load lines $I_{MTJ} = (V_{CELL} - V_{MTJ})/R_{MTJ}$ correspond to the MTJ resistance in both the P state and the AP state. The intersects of the load line to the transistor $I_{ds} - V_{ds}$ (at $V_{gs} = V_{dd}$) are the cell currents; V_{MTJ} is the voltage from $V_{MTJ} = V_{CELL} - V_{ds}$ at the intersection. Since there are two MTJ states, there are two MTJ load lines and two cell currents, I_H and I_L. Note that, when the MTJ changes state, the voltage across the MTJ also changes. Thus, the ratio of the cell current I_H/I_L is smaller than the ratio of R_{AP}/R_P. When the MTJ switches from the AP to the P state, as shown in Fig. 6.23(d) at step 3, the cell current rises and V_{MTJ} decreases, while the voltage across the transistor increases by the same amount. Conversely, during the write

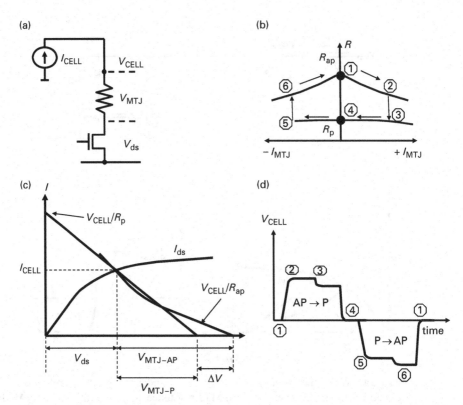

Figure 6.24. Write operation of a STT MRAM cell under constant-current write condition: (a) equivalent circuit schematics; (b) MTJ resistance switching trajectory; (c) MTJ load lines for determining the cell voltages; (d) voltage waveform when writing AP-to-P and P-to-AP.

operation from the P to the AP state, a positive voltage is applied to the BLC, and BLT is ground.

When the cell is under a constant-current bias, the transistor drain-to-source voltage does not change; only the MTJ bias voltage, V_{MTJ}, changes when it switches states. Thus, V_{CELL} changes as shown in Fig. 6.24.

6.9.2   Read and write voltage window of STT memory cell

This section deals with the read and write operation margin of a 1T-1MTJ cell when it is operated in STT mode. To do this, we need to examine the statistical properties of the MTJ. Both the read voltage and the write voltage are bounded by the MTJ's intrinsic properties and the distribution of the device parameters [30].

The read and write voltage margin of a STT cell array is illustrated in Fig. 6.25. The write voltage, V_W, is bounded on the upper side by the distribution of the

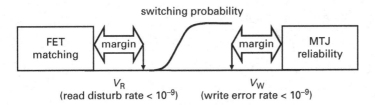

switching probability

| FET matching | margin | | margin | MTJ reliability |

V_R
(read disturb rate < 10^{-9}) V_W
(write error rate < 10^{-9})

Figure 6.25. Read/write operational voltage window of the STT MRAM cell.

MTJ tunnel barrier breakdown voltage and on the lower side by the write error rate. The read voltage, V_R, is bounded on the upper side by the read disturb rate and on the lower side by the amplitude of the sense signal. The sense signal should be greater than the mismatch of the sense amplifier to warrant accurate data sensing. The read disturb rate and the write error rate must be sufficiently low for practical RAM application, at least below 10^{-9}.

Since the MTJ junction breakdown and the sense amplifier mismatch set the maximum allowable range of $V_W - V_R$, the bias voltage dependence of the write error rate and the read disturb rate should be as abrupt as possible. Based on Eq. (6.9), the upper limit of V_W sets the shortest write pulse t_{PW} and/or the upper limit of RA; $V_W - V_R$ sets the smallest switching-energy barrier, E_b. The lower limit of E_b is also set by the requirement of the retention time.

6.9.3   Sense signal margin

During the read operation, the read voltage is less than 0.15 V. At such a low V_{MTJ}, the transistor is operating in the linear regime, and therefore it can be treated as a linear resistor. The cell resistance is the sum of R_{MTJ} and R_{ds} of the MOS transistor:

$$R_{CELL} = R_{MTJ} + R_{ds}. \qquad (6.17)$$

The statistical variation of R_{CELL} must take into account the variations of both the MTJ and the transistor as

$$\sigma(R_{CELL}) = [\sigma(R_{MTJ})^2 + \sigma(R_{ds})^2]^{1/2}. \qquad (6.18)$$

When the MTJ is in the P state, the cell resistance is R_{CELL-L}. In a large memory array, the value of R_{CELL-L} varies around its mean value, $\langle R_{CELL-L} \rangle$. Similarly, when the MTJ is in the AP state, the cell resistance is R_{CELL-H}. The tail ends of the R_{CELL-L} and R_{CELL-H} cell should not overlap, and the separation must be larger than the mismatch of the sense amplifier. Thus,

$$\text{read margin} = \frac{\langle R_{CELL-H} \rangle - n \cdot \sigma(R_{CELL})}{\langle R_{CELL-L} \rangle + n \cdot \sigma(R_{CELL})} - 1, \qquad (6.19)$$

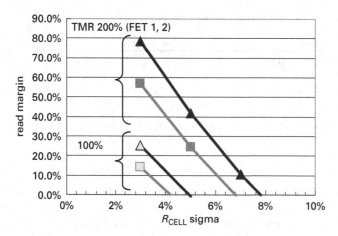

Figure 6.Q6. Read signal margin vs. the sigma of cell resistance. Two transistor sizes are considered for each TMR.

where n is a function of the number of bits in the memory. Note that, since $TMR = \frac{\langle R_{AP} \rangle}{\langle R_P \rangle} - 1$, the read margin is reduced by the source-to-drain resistance of the MOS transistor and the σ of the total cell resistance.

Homework

Q6.6 Find the read margin of a 64 Mb STT memory chip having 1T-1MTJ cell. The MTJ $R_P = 1000\ \Omega$ and the TMR ratio is 200%; try two transistors with drain-to-source resistance $R_{ds} = 1360\ \Omega$ and $680\ \Omega$. The distribution of R_{CELL} is Normal and the range $\sigma(R_{CELL}) = 3\%$, 5% and 7%. Try another case with MTJ TMR $= 100\%$.

A6.6 First, we find how many σ one must consider to cover the entire distribution of cell resistance in a 64 Mb array. From Appendix D, we find that $n \sim 5.5$ for a 64 Mb chip. Based on the read margin from Eq. (6.19), one can construct Fig. 6Q.6. The gate width of the MOS transistor FET1 is larger than that of FET2. Its drain-to-source resistance is smaller, thus the effective TMR ratio of the cell is larger. The cell can tolerate larger variations in cell resistance.

6.9.4   Write-to-breakdown-voltage margin

The intrinsic dielectric breakdown voltage of a MTJ tunnel barrier is proportional to the electric field in the tunnel barrier, $E = V/d$, where V is the voltage across the tunnel barrier and d is the thickness of the tunnel barrier. The tunnel current is inversely proportional to the logarithmic function of d. Thus, the write voltage is proportional to RA, while the breakdown voltage is proportional to $\ln(RA)$. Thus, one way of increasing the separation between the write voltage and the breakdown voltage is to design a MTJ with a lower RA. The only drawback is that a low-RA tunnel barrier is physically thinner, and therefore is more likely to have a higher number of defects. Most of the soft breakdowns have been attributed to the defects in the tunnel barrier.

6.10   Data retention and E_b extraction method

The switching-energy barrier E_b is the most important parameter in determining the data retention time. There are many ways to extract E_U; the most popular way is based on Eq. (6.8). We only need to extract the slope of the switching current from a set of write pulse width on a semi-log plot. However, since the switching current heats up the MTJ, the MTJ is at different temperatures when measured at different write currents. Thus, the extraction is affected by the change of temperature.

The switching-energy barrier E_b can also be extracted using a time-dependent switching under a field method at constant temperature based on a *magnetization decay model* [31]. This method does not involve writing a MTJ with current, rather it examines the time-to-magnetic reversal of a large number of MTJs under a magnetic field below the switching threshold. It is equivalent to the accelerated test of time-to-dielectric-breakdown for extracting the life expectancy of a dielectric. The E_b extracted from this method corresponds to the data retention time when MTJ is idling. The magnetization decay model is described as follows.

The switching-energy barrier when reduced under an external field is given by

$$\Delta = \Delta_0 (1 - H/H_{C0})^2,$$

where $\Delta_0 = E_b/k_B T$. The thermal agitation introduces a finite probability of magnetization reversal, which can be expressed as follows:

$$s = f \cdot \exp(-E_b/k_B T), \tag{6.20}$$

where f is the attempted frequency, which is of the order of 10^9 s^{-1}. By measuring the accumulative function of time-to-switching t of a large number of MTJs, one can obtain the *reduced magnetization, m*, as follows:

$$m(t) = 2 \cdot \exp(-st) - 1. \tag{6.21}$$

Initially, at time zero,

$$m(t = 0) = 2 - 1 = 1,$$

meaning all MTJs are initialized to, say, the AP state. Under a field, MTJs will gradually switch, some earlier and some later. Eventually,

$$m(t = \infty) = -1,$$

and all the MTJs switch to the P state.

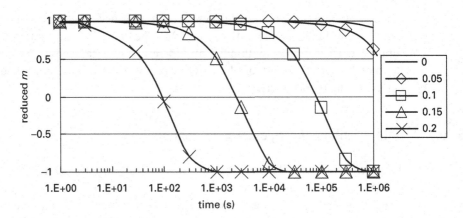

Figure 6.26. Reduced magnetization m vs. time as a function of external field H/H_{C0} from 0 to 0.2, step 0.05. Thermal factor $\Delta = 40$. During the write operation, the switching-energy barrier is reduced to ~0 by the injection of a large write current I_W.

For each external field, the fractions of total MTJs that switch polarity over a series of time points are recorded. The time point at which 50% of the MTJs switch is called t_C. At this point, $m(t_C) = 0$ and $H = H_C$. By fitting this curve, both H_{C0} and E_b can be extracted:

$$H_C(t_C) = H_{C0}(1 - [(k_B T/E_b) \ln(s t_C/C)])^{0.5}. \tag{6.22}$$

where $C = -\ln(0.5) = 0.31$.

The E_b extracted from this measurement is when the sample is at an ambient temperature. It has been reported that the E_b extracted with this method is much greater than that extracted from the write pulse width measurement.[*] See Fig. 6.26.

6.11   Thermal stability of STT memory chip

This section concerns the thermal stability of STT cells or data retention. In the STT array, each cell is written with a current that is fed individually. The cell array does not suffer from the half-select disturbance problem as in the field-write MRAM cells. Rather, it may be disturbed when the cell is read. There is a 50% chance that the read current reduces the switching-energy barrier, depending on the direction of the read current and the state of the cell.

Consider the energy barrier E_b between the two states of the MTJ, namely the polarization of the free layer is parallel or antiparallel to the pinned layer.

[*]Courtesy of MagIC Technologies, Inc.

During the write operation, the switching-energy barrier is reduced to ~0 by the injection of a large write current I_W. During the read operation, the switching-energy barrier is reduced to $C \cdot E_b$, where $C = (1 - I_R/I_{C0}) < 1$, where I_R is the read current. Figure 6.26 illustrates the energy at these three conditions.

The probability of one error occurring in a STT MRAM chip over a period of time is defined as follows:

$$\sum_{i=1}^{\text{bits}} \sum_{j=1}^{3} P_{ij}\tau_{ij} = 1, \tag{6.23}$$

where i and j are the number of bits and the state of the bit cell ($j = 1$: stand-by, 2: read, 3: write), respectively; P_{ij} is the probability of magnetization reversal of cell i in state j; similarly, τ_{ij} is the time duration of cell I in state j. Thus,

$$\sum_{i=1}^{\text{bits}} (P_{is}\tau_{is} + P_{ir}\tau_{ir} + P_{iw}\tau_{iw}) = 1, \tag{6.24}$$

where subscript s is for stand-by, r is for read and w is for write.

Homework

Q6.7 Derive the switching-energy barrier required for a 1 Mb STT MRAM chip with the following assumptions: (1) read current $= 0.3I_{C0}$ (refer to the switching energy barrier diagram in Fig. Q6.7); (2) the read access occurs five times as often as the write access, the chip is accessed continuously, and the write pulse width and the read pulse width are the same, both lasting half of the access cycle; (3) 8 bits out of the total 1 Mb of the chip are accessed all the time; (4) no bit fail in 10 years.

A6.7 Note that K bits of the N-bit array are accessed each time. There are $(N - K)$ cells in the stand-by condition all the time. There are N cells with 1/6 of the total time period being written, and 5/6 being read. Since $\tau_{ir} = 5\tau_{iw}$, over a period of t_{total}, $\tau_{is} = t_{\text{total}}$, $\tau_{ir} = 5/6\, t_{\text{total}}$. From Eq. (6.9) the switching probability of the MTJ being in stand-by is given by

$$P_{\text{SW-s}} = 1 - \exp\left\{-\frac{t_S}{\tau_0}\exp\left[-\frac{E_b}{k_BT}\right]\right\},$$

and of those being read is

$$P_{\text{SW-r}} = 1 - \exp\left\{-\frac{t_R}{\tau_0}\exp\left[-\frac{E_b}{k_BT}\left(1 - \frac{I}{I_{C0}}\right)\right]\right\}.$$

The stability of the cell during the time taken to write, τ_{iw}, is not a concern. The equation can therefore be written as

$$(N - K) \cdot P_{\text{SW-s}}\, t_s + KP_{\text{SW-r}}\, t_r = 1.$$

For 10 years, we use $t_{\text{total}} = 3.15 \times 10^8$ s. Since the MTJ is in one of two states (AP or P), there is a 50% chance that the read current flows in the direction

(a)   (b)

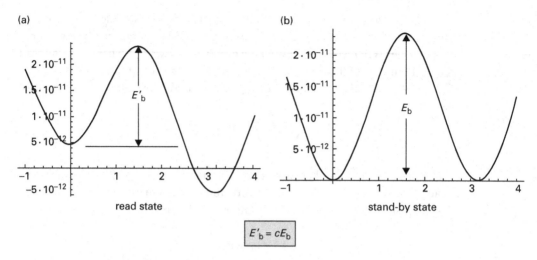

read state   stand-by state

$$E'_b = cE_b$$

Figure 6.Q7. Schematics of switching energy diagram in (a) read states and (b) stand-by.

that lowers the energy barrier of the MTJ. The read current pulse width is half the read cycle time. Thus, the total read pulse time over 10 years is

$$t_r = 3.15 \times 10^8 \times \frac{5}{7} \times \frac{1}{2} \times \frac{1}{2} = 5.63 \times 10^7.$$

For a 1 Mb chip there are $N = 2^{20}$ cells, and eight cells are selected, so $K = 8$. The switching-energy barrier during read is $E'_b = (1 - 0.3)E_b$. One can obtain $E_b = 58.57\ k_BT$ for a 1 Mb chip.

Q6.8 Assume that a MRAM chip is operated under the following conditions: (1) 8 bits out of the total 1 Mb of the chip are accessed all the time; (2) the frequency of the read access is as often as that of the write access; (3) the write pulse width and the read pulse width are the same, both lasting half of the access cycle. Derive the switching-energy barrier required for

(a) no fail bits in 10 years for a 1 Mb–1 Gb STT MRAM chip, and the read current $= 0.1I_{C0}$;
(b) no fail in 10 years for a 1 Mb chip with read current in the range $0.1–0.5I_{C0}$;
(c) no fail over a range in time from 1 s to 10 years.

A6.8 Refer to Figs. 6.Q8(a)–(c).

6.12   Write-current reduction

To reduce the write current, one needs to work on the material properties, such as the reduction of the damping constant and the M_St of the free layer, where t is the thickness of the free layer. These may be discovered during research of the film

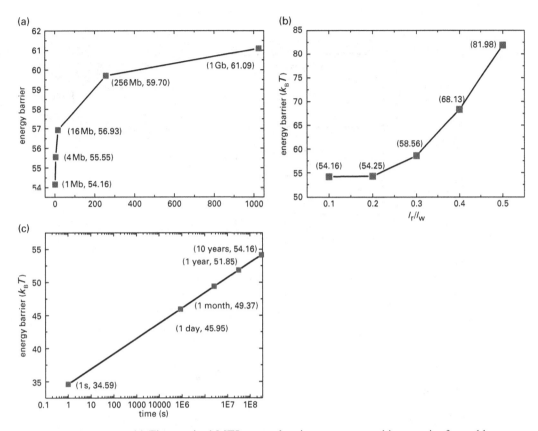

Figure 6.Q8. (a) The required MTJ energy barrier vs. memory chip capacity for stable operation. (b) The required MTJ energy barrier vs. read current of a 1 Mb chip. (c) The required MTJ energy barrier vs. operation time.

material. In addition, one may need to raise the temperature of the MTJ during the write operation. This section introduces two proposed film-stack structures for the longitudinal MTJ, in which the magnetization is in the film plane, and a structure for a perpendicular MTJ, in which the magnetization is normal to the film plane.

6.12.1   Nanocurrent-channel film-stack structure

One way to reduce the switching current density is to constrict the spin polarized current into small current channels in the free layer of the MTJ. The free layer is made up of three layers: a thin layer of SiO (insulator), doped with a minute amount of Fe [32], set between two layers of CoFe. The Fe in the SiO layer forms conducting channels between the two CoFe layers. A practical volume fraction of Fe is of the order of ~10%. The current through the free layer is constricted through the Fe conducting channels at the SiO layer. Such a channel is called a

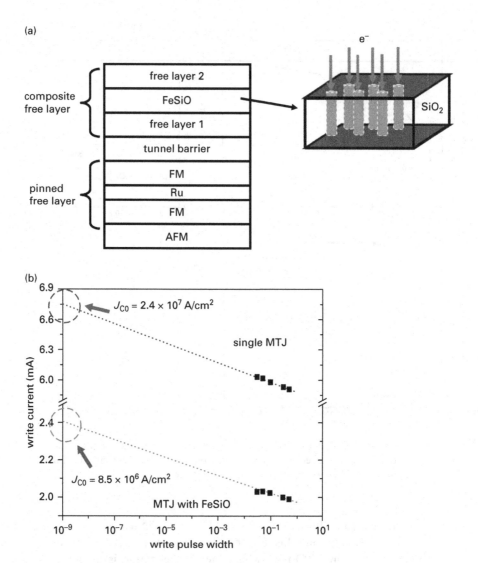

Figure 6.27. (a) Free layer structure of nanocurrent-channel MTJ. (b) Write current. (After ref. [32], with permission from AIP.)

nanocurrent channel. The local current density in the channel is orders of magnitude greater than the average current in the CoFe, and the temperature at the channel is higher than the rest of the free layer. Compared to a single FM free layer of the same size, the switching current is ∼2.5 times smaller. It is believed that the combination of the local high current density and higher temperature at the channel triggers the magnetic reversal at the channel and spreads from the channels to the entire free layer. Figure 6.27 shows the switching current. Since the total volume of the CoFe material is the same, the thermal stability of the MTJ is unchanged.

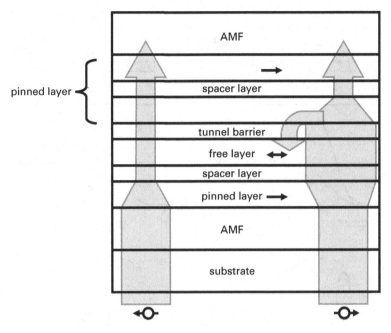

Figure 6.28. Double-spin-filter MTJ film stack and polarized current flow.

6.12.2   Double-spin-filter structure

Another method used to reduce the write current is to improve the efficiency of the spin-torque transfer from polarized free electrons to act on the free layer. Figure 6.28 shows a double-spin-filter MTJ structure [33]. Pinned layers are placed near the free layer: one on each side, with a spacer between them and the free layer. The magnetizations of these two pinned layers are in opposite directions. Electrons coming from the bottom of the structure are polarized (filtered) by the bottom pinned layer. After they leave the free layer, they are reflected from the top pinned layer. The passing electrons and reflected electrons both act on the free layer. Thus, the torque transfer efficiency is improved. A very low J_{C0} has been reported, of the order of 10^6 A/cm^2 [34].

6.12.3   Perpendicular MTJ

Until now, we have discussed the properties of MTJs of a special type: the longitudinal MTJ (L–MTJ). The anisotropy of the free layer and pinned layer of a L–MTJ is in the film plane. In this case, the demagnetization field plays a dominant role in the switching process of the free layer of the MTJ. In the field MRAM case, the external field rotates the magnetization in the film plane against the in-plane anisotropy. In the STT MRAM, the polarized electron current excites the magnetization to process against the damping, and the precession is hindered

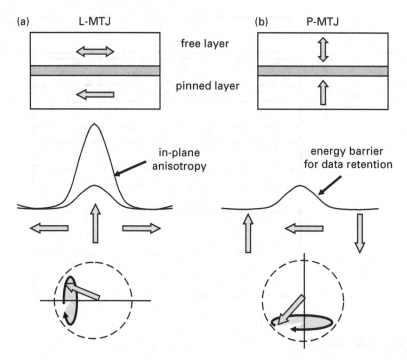

Figure 6.29. Magnetization of the film stack, switching-energy barrier, precession axis of (a) L-MTJ and (b) P-MTJ. (After ref. [35], with permission from IEEE.)

by the out-of-plane anisotropy. Since the film thickness is much smaller than its lateral dimension, the out-of-plane anisotropy is much greater than the in-plane anisotropy.

One way to reduce the switching current is to make a MTJ with a perpendicular crystalline anisotropy, namely the crystalline anisotropy is out-of-plane. In the perpendicular MTJ (P-MTJ) the magnetizations of the free layer and the pinned layer are out of the film plane. Thus, the magnetization is along the out-of-plane direction when idling. When excited by the polarized current, the precession of the magnetization is no longer hindered by the out-of-plane anisotropy. As a result, it takes less torque (thus, energy) to switch the magnetization. For the same thermal energy barrier, the switching-energy barrier of the L-MTJ is much larger. Figure 6.29 illustrates the difference in precession process and the switching-energy barrier between the L-MTJ and the P-MTJ.

There are a few interesting properties of the P-MTJ. (1) It is free of edge curling, since the magnetization is out-of-plane. Since the magnetization does not point toward the MTJ edge, a vortex does not form, even the MTJ is in a circular shape; thus, it is more stable. The aspect ratio of the L-MTJ must be greater than 2 to prevent the formation of a vortex in the free layer. (2) It can have a very high out-of-plane anisotropy ($H_K > 1$ kOe) without affecting the switching current. The high crystalline anisotropy energy allows the P-MTJ to store data for longer times

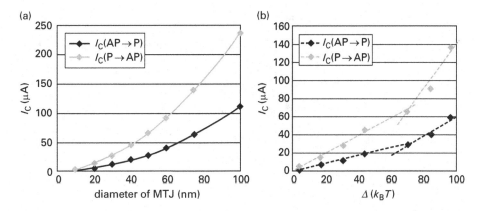

Figure 6.30. Simulated switching current vs. (a) diameter of the P-MTJ, (b) switching-energy barrier of P-MTJ. (After ref. [35], with permission from IEEE.)

with a smaller free layer volume; thus, it can be scaled beyond the dimension of a stable L-MTJ.

Micromagnetic study indicates that a P-MTJ with a damping constant α of 0.03–0.05 and a size of around 50 nm can simultaneously satisfy the requirement of programming current less than 100 μA and 10 year data retention ($\Delta = 60\,k_B T$). The dependence of the programming current on retention energy is shown in Fig. 6.30. The slope corresponds to the efficiency of the spin transfer. The switching-energy barrier Δ does not have a simple dependence on the diameter of the P-MTJ element. In the case of $\Delta < 60\,k_B T$, the slope is smaller, corresponding to coherent switching. For $\Delta > 60\,k_B T$, the switching is incoherent [35].

A recently developed P-MTJ consists of the following: under-layer/L1$_0$-alloy free layer/MgO/perpendicular reference layer/capping layer, in which an L1$_0$-crystalline ordered alloy like FePt has been implemented in the P-MTJ [36]. It has a large anisotropic energy K_u of the order of 10^7 erg/cc (about one order greater than that of a typical L-MTJ) and a high thermal stability of $60\,k_B T$ [37–40].

Many film structures have been proposed that involve the addition of a perpendicular component in the longitudinal anisotropy free layer to aid the growth of the precession cone angle to lower the switching current density. One way of achieving this is to leave the free layer anisotropy in-plane whereas the pinned layer is out-of-plane [41]. Other ways involve tapering the wall edge of the MTJ film stack to induce a perpendicular-component-shape anisotropy to trigger switching at lower current density [14].

6.13 Direct observation of magnetization reversal

An X-ray beam was applied to detect local magnetization vector in a GMR structure with fine spatial resolution [42]. The reconstructed magnetization of

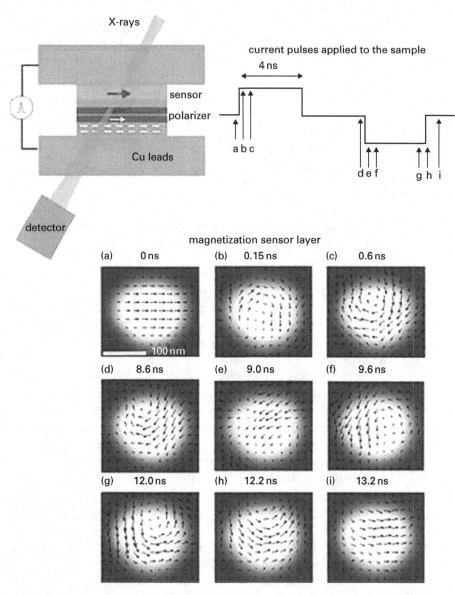

Figure 6.31. (a) X-ray direct observation of local magnetization in a 150×100 nm^2 GMR structure under high current injection. (b) The local magnetization vector at nine time instants. (After ref. [41].)

the free layer at nine time instants in the positive and negative write current cycle are shown in Fig. 6.31. The Oersted field from the write current is estimated to be in the order of 100 Oe around the perimeter of the structure, for a current density of mid-10^7 A/cm^2 [43]. It is evident that the local magnetization exhibits a C-state during the write pulse, which results from the strong circular magnetic field of the high write current density. A vortex is formed when the write current is higher. The sense of the vortex corresponds to the direction of the current. There is a lot of interplay between the field and the spin torque of the electron flux during the switching. If this vortex remains after the write current pulse is terminated, the GMR switches into an intermediate state. The probability of forming a permanent vortex is lower in an MTJ when the aspect ratio of the MTJ shape is greater.

References

[1] J. C. Slonsczewski, *Electronic Device Using Magnetic Components*, US Patent 5695894 (1997).

[2] Y. Lassailly, H.-J. Drouhin, A. J. van der Sluijs, G. Lampel and C. Marliere, *Phys. Rev. B* **50**, 13054 (1994).

[3] G. J. Strijkers, *Phys. Rev. B* **63**, 104510 (2001).

[4] S. X. Huang, Y. T. Chen and C. L. Chien *Appl. Phys. Lett.* **92**, 242509 (2008).

[5] J. C. Slonczewski, *J. Magn. & Magn. Mater.* **159**, L1 (1996).

[6] L. Berger, *Phys. Rev. B* **54**, 9353 (1996).

[7] M. Tsoi, A. G. M. Jansen, J. Bass, W.-C. Chiang, M. Seck, V. Tsoi and P. Wyder, *Phys. Rev. Lett.* **80**, 4281 (1998).

[8] E. B. Meyers, D. C. Ralph, J. A. Katine, R. N. Louie and R. A. Buhrman, *Science* **285**(6), 867 (1999), www.sciencemag.org.

[9] J. Grollier, V. Cros, A. Hamzic *et al.*, *Appl. Phys. Lett.* **78**(23), 509 (2001).

[10] D. C. Ralph and M. D. Stiles, *Spin Transfer Torques*, http://cnst.nist.gov/epg/Pubs/pdf/ epg791.pdf (2007).

[11] A. Fert, V. Cros and J.-M. George *et al.*, *J. Magn. & Magn. Mater.* **272–276**(3), 1706 (2004).

[12] J. Z. Sun, *Phys. Rev. B* **62**, 570 (2000).

[13] I. N. Krivorotov, D. V. Berkov, N. L. Gorn, N. C. Emley, J. C. Sankey, D. C. Ralph and R. A. Buhrman, *Phys. Rev. B* **76**, 024418 (2007).

[14] P. M. Braganca, I. N. Krivorotov, O. Ozatay, A. G. F. Garcia, N. C. Emley, J. C. Sankey, D. C. Ralph and R. A. Buhrman, *Phys. Rev. B* **77**, 144423 (2008).

[15] J. C. Slonczewski, *J. Magn. & Magn. Mater.* **195**, L261 (1999).

[16] R. H. Koch, J. A. Katine and J. Z. Sun, *Phys. Rev. Lett.* **92**(8), 088302 (2004).

[17] M. Hosomi, H. Yamagishi, T. Yamamoto, *et al.*, *IEDM Technical Digest*, **459** (2005).

[18] T. Aoki, Y. Ando, D. Watanabe, M. Oogane and T. Miyazaki, *J. Appl. Phys.* **103**, 103911 (2008).

[19] K. Yagame, M. Hosomi, I. Olmori, T. Yamamoto, Y. Higo, Y. Oishi and H. Kano, *J. Appl. Phys.* **97**, 10C707 (2005).

[20] Y. Higo, K. Yamane, K. Ohba, H. Narisawa, K. Bessho, M. Hosomi and H. Kano, *Appl. Phys. Lett.* **87**, 082502 (2005).

[21] T. Inokuchi, H. Sugiuama, Y. Saito and K. Inomata, *Appl. Phys. Lett.* **89**, 102502 (2006).

[22] T. Min, Q. Chen, T. Torng, C. Horng, D. Tang and P. Wang, *J. Appl. Phys.* **105**, 07C931 (2009).

[23] J. Das, R. Degraeve, P. Roussel, G. Groeseneken, G. Borghs and J. De Boeck, *J. Appl. Phys.* **91**(10), 7712 (2002).

[24] S. Bae, J. H. Judy, I. -F. Tsu and M. Davis, *J. Appl. Phys.* **94**(12), 7636 (2003).

[25] J. Akerman, P. Brown, M. DeHerrera *et al.*, *IEEE Trans., Device & Materials Reliability* **4**(3), 428 (2004).

[26] S. C. Li, J. -M. Lee, M. -F. Shu, J. P. Su and T. -H. Wu, *IEEE Trans., Magnetics* **41**(2), 899 (2005).

[27] J. G. Simmons, *J. Appl. Phys.* **34**, 1793 (1963).

[28] G. Landry, Y. Dong, J. Du, X. Xiang and J. Q. Xiao, *Appl. Phys. Lett.* **78**(4), 501 (2001).

[29] P. Padhan, P. LeClair, A. Gupta, K. Tsunekawa and D. D. Djayaprawira, *Appl. Phys. Lett.* **90**, 142105 (2007).

[30] R. Beach, T. Min, C. Horng *et al.*, *IEDM Technical Digest* (2008), paper 12.5.

[31] M. P. Sharrock, *J. Appl. Phys.* **76** (10), 6413 (1994).

[32] H. Meng and J. P. Wang, *Appl. Phys. Lett.* **89**, 152509 (2006).

[33] Y. Huai and P. Nguyen "Magnetic element utilizing spin transfer and MRAM devices using the magnetic element," US Patent 6,714,444 (2004).

[34] Z. Diao, A. Panchula, Y. Ding *et al.*, *Appl. Phys. Lett.* **90**, 132508 (2007).

[35] T. Kishi, H. Yoda, T. Kai, *et al.*, *IEDM Technical Digest* (2008), paper 12.6.

[36] M. Hagiudaa, S. Mitania, T. Sekia, K. Yakushijia, T. Shimab and K. Takanashia, *J. Magn. & Magn. Mater.* **310**(2), 1905 (2007).

[37] T. Nagase, K. Nishiyama, M. Nakayama, N. Shimomura, M. Amano, T. Kishi and H. Yoda, APS March Meeting, New Orleans, March 10–14 (2008).

[38] M. Nakayama, T. Kai, N. Shimomura *et al.*, *J. Appl. Phys.* **103**, 07A710 (2008).

[39] H. Yoda, T. Kishi, T. Nagase *et al.*, *Intermag 2008 Digest, FA-04* (2008).

[40] M. Yoshikawa, T. Kai, M. Amano, *et al.*, *Intermag 2008 Digest, AC-01* (2008).

[41] A. Kent, and D. Stein, "High speed low power annular magnetic devices based on current induced spin-momentum transfer," US Patent 7,307,876 (2007).

[42] J. Stohr, *Ultrafast Magnetic Switching of Nanoelements with Spin Currents*, http://www.ssrl.slac.stanford.edu/stohr/spininjection.htm (2007).

[43] J. P. Strachan, V. Chembrolu, Y. Acremann *et al.*, *Phys. Rev. Lett.* **100**, 247201 (2008).

7 Applications of MTJ-based technology

7.1 Introduction

A short time after the discovery of magnetic tunneling devices, tunneling magneto-resistance (TMR) replaced giant magnetoresistance (GMR) read sensor in the hard disk drive. This marked the first successful commercialization of magnetic tunnel junction technology. The first mass production of the magnetic recording head based on a MgO tunnel barrier took place in 2006. In the same year, 4 Mb MRAM chips were commercialized, and it was the first field-MRAM product working in the toggle-write mode. The viability of MTJ technology at the product level is proven. Subsequently, electronic system designers started to consider seriously how to take advantage of this technology. Many new applications of MTJ technology begin to emerge. One of the new circuit elements is the non-volatile magnetic flip-flop device, which is used for the reduction of VLSI chip power as well as for run-time system re-configuration. Such new applications can only be realized with the unique properties of magnetic tunnel junction devices.

Other new applications are being explored in the field of healthcare [1–3]. GMR and TMR chips are used for detecting biological molecules labeled with magnetic particles, and this could be a powerful platform for next-generation diagnostics. The sensitivity achievable with simple portable instrumentation can be orders of magnitude better than the current methods. Since this application is still in its infancy, it will not be discussed further here. Interested readers are referred to the references above.

First, we will look into the memory landscape and show the competitive position of the MRAM. Then, we will discuss the application of MTJs in the CMOS system-on-chip (SoC) as a non-volatile switching element.

7.2 MRAM market position

The random access memory market of today can be summarized in Fig. 7.1. The mainstream semiconductor memory products under mass production are SRAM, DRAM (both are volatile memory) and the NOR-type, NAND-type FLASH memory (both are non-volatile memory). Large-capacity memories are used for storage, such as NAND-type FLASH and a non-semiconductor product,

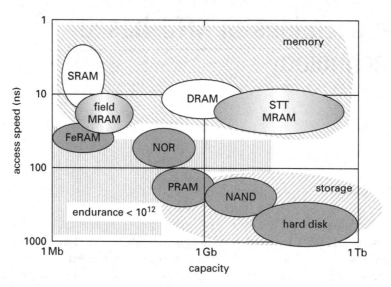

Figure 7.1. Memory market landscape.

such as a hard-disk drive (HDD). The memory landscape, in general, is characterized by the access speed and cost. The cost is inversely related to the cell density at a given CMOS manufacturing node. In 2007, the most advanced memory manufacturing node was at 50 nm. A large-capacity memory chip is usually made of higher-density memory bit cells. The density of a NAND-type FLASH bit cell can be larger than the physical cell at a given CMOS node, since each physical cell may store more than 1 bit.

Although the NAND-FLASH and the NOR-FLASH are non-volatile, their application is restricted. Their write endurance is less than 10^6 time, meaning that they cannot be written more than one million times. Although their read access is very fast, their write access is very slow, of the order of milliseconds. Thus, they are only suitable for data storage. On the other hand, SRAM and DRAM are the two memories with characteristics of infinite write endurance and fast access. The modern logic processor is so fast that only SRAM can keep up with the pace of data demand from the processor. DRAM is frequently used as level-two (L2) memory in a system. Other than SRAM, all other types of memory require a memory controller to manage the read, write and refresh operations. SRAM is designed to interface directly with the processor, or central processing unit (CPU).

The industry has been looking for a unified memory that can read/write like SRAM and is non-volatile and dense like NAND-FLASH and has infinite endurance like SRAM. The dream memory is called "universal memory," which, as yet, is not commercially available.

There are a few new memory candidates that intend to fill the role of universal memory. Among the contenders, ferroelectric memory (FeRAM), phase-change memory (PRAM) and magnetic memory (MRAM) are the stronger candidates. The resistive RAM (RRAM) is in the early study stage, and is much less mature than the first three. Their write access cycle time is comparable and short (relative to FLASH), and they are all non-volatile. At the time of writing, FeRAM and PRAM both show write endurance of the order of 10^9–10^{12} cycles, much better than FLASH. Nonetheless, their write endurance cycle is still limited. MRAM is the only one with unlimited endurance and read/write like SRAM. The read operation of FeRAM is destructive, meaning that the data must be restored after read, similar to the DRAM operation.

FeRAM and MRAM are currently under manufacture. The MRAM in production is field MRAM, the cell size of which is larger than that of the STT MRAM. Between longitudinal and perpendicular MTJ-based MRAM, the former is likely to be manufactured first. Although the perpendicular MTJ is more scalable than the longitudinal MTJ, the technology is less mature. With good cell density and scalability as in a DRAM cell, fast read/write access like SRAM, along with non-volatile storage and infinite write endurance, the perpendicular STT MRAM is closer to a "universal memory."

The field MRAM occupies the lower-density SRAM market, whereas the STT MRAM is being developed to address the higher-density DRAM market. Since the read-write characteristics of MRAM are very close to an SRAM, or, more precisely, a non-volatile SRAM, its first application has been the replacement of the combination of SRAM and a NOR-FLASH. As a result, the 4 Mb MRAM chips on the market have an asynchronous SRAM interface. The read (or write) access time is of the order of tens of nanoseconds. Such a short access time cannot be offered by any FLASH devices. (It takes milliseconds to write data into FLASH.) The fastest MRAM experimental 32 Mb chip can be accessed in a 12 ns clock cycle, read or write [4].

Current MRAM chips are designed with an asynchronous SRAM interface. MRAM chips with greater capacity are being designed to replace the combination of DRAM and NOR-FLASH. Thus, the interface may be asynchronous and synchronous with burst or page mode to increase data bandwidth. Many DRAM chips today are operated in synchronous mode or page mode to provide higher data throughput [5].

All MRAM chips are designed with a data protection feature, which is not applicable to SRAM. Since MRAM chips retain the on-board data while the chip power is turned off, one must prevent "accidental write" of the chip when the chip is at the power-up stage. During the power-up stage, the chip is automatically set into "write protect" mode to protect the on-board data. Only after the chip receives a special "unlock" instruction from the memory controller does the chip start functioning normally. The unlock instruction, for example, can be a series of read access at specific addresses. After the MRAM receives this instruction, the chip starts working normally, like a SRAM.

A relative performance comparison of the different types of memory is given in Table 7.1.

Table 7.1 Memory property comparison

	SRAM	DRAM	NOR	NAND	FeRAM	Field MRAM	PRAM	STT MRAM
Data retention	no	no	yes	yes	yes	yes	yes	yes
Cell size (F2)	80–120	6–12	10	2–5	15–35	35–45	6–12	6–20
Read access (ns)	1–50	30	10	50	20–80	3–20	20–50	2–20
Write/erase time (ns)	1–50	50	$10^5/10^7$	$10^6/10^5$	50	3–20	50/200	2–20
Rewrite endurance	$>10^{16}$	$>10^{16}$	10^5	10^5	10^{9-12}	10^{15}	10^{9-12}	$>10^{15}$
Power at write	low	low	high	high	low	high	low	low
Power other than read, write, erase	leakage current	refresh current	none	none	none	none	none	none
Input high voltage (V)	no	2	6–8	16–20	2–3	3	3.0	no
Product availability	yes	yes	yes	yes	yes	yes	prototype	prototype

Source: Digest of ISSCC websites.

7.3   MTJ applications in CMOS SoC chips

The silicon wafer process has evolved into two major categories: logic process and memory process (usually referred to as DRAM or FLASH process). The logic process provides a single layer of polysilicon and three to ten layers of metal. The number of metal layers increases with the complexity of the SoC chip. The memory process provides up to two or three layers of polysilicon and two or three layers of metal interconnect. The processing cost of the logic chip is usually higher than that of the memory chip, since the cost of the metal layer is higher than that of the polysilicon layer. All stand-alone MRAM products are made with the CMOS logic process, not with the memory process, for the reason that the polysilicon layer is too resistive and cannot support the large write current. Thus, the complexity of MRAM wafer process is the same as the CMOS logic process.

7.3.1   Embedded memory in logic chips

One way to improve the overall performance of a SoC chip is to embed a high-speed memory in the SoC. Then, the CPU first accesses data from the embedded memory rather than from the external memory chip. Only when data are not found in the embedded memory (called a miss) does the CPU access data from the external memory chip. As long as the hit rate of finding data from the embedded memory is high, the CPU performance is improved, due to the shorter access path and thus the delay time. Frequently, two levels of memory are embedded in SoC chips: level-1 (L1) is ultra-fast SRAM and level-2 (L2) is dense DRAM.

For this application, the access time of the embedded MRAM is designed to be as short as possible. (Readers are referred to Fig. 5.30 of ref. [6].) The memory cell is designed to provide a large sense signal and the write current is localized to one cell to prevent write disturbance to the half-selected cells. The access time of the chip usually comes at the expense of the cell size: 2T-1MTJ or 5T-2MTJ cells have been considered; the cell density of such cells is close to, but smaller than, that of the 6T-SRAM cell.

7.3.2   Unbalanced MTJ flip-flop

Magnetic tunnel junction devices are used as *switching binary resistors* in a new class of circuits based on an unbalanced MTJ flip-flop logic circuit. The commercial name for these is *magnetic flip-flop devices* [7]. A MTJ flip-flop is non-volatile, and we will discuss two of its derivatives, such as the multiplexer and the data register.

Most switching elements in a logic circuit are required to switch with a sub-nanosecond gate delay and for an unlimited number of times. A MTJ is the only non-volatile element that can meet these two requirements. In addition, MTJ technology scales well with CMOS technology. Thus, the magnetic flip-flop will be used in present and future generations of CMOS technology.

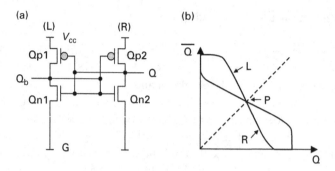

Figure 7.2. (a) SRAM-based flip-flop. (b) Transfer curve and metastable point.

The SRAM-based flip-flop is a balanced flip-flop, as long as the two inverters are identical. The flip-flop is made of two cross-coupled inverters: the output of one is connected to the input of the other, as shown in Fig. 7.2(a); Fig. 7.2 (b) shows the inverter input–output transfer curves of these two inverters. The flip-flop has one metastable state at the intercept point P of the two transfer curves. There are two stable states: $(Q, Q_b) = (V_{cc}, 0)$ and $(0, V_{cc})$, and they are of equal distance from P. When the flip-flop is powered up, it settles into either of the two stable states with equal probability (as long as the inverters are truly identical).

The unbalanced MTJ flip-flop is shown in Fig. 7.3(a). MTJ acts as a source follower to the NMOS transistors. When the two MTJs are programmed into complementary states, the two inverters show different transfer curves, and the flip-flop is therefore an unbalanced flip-flop.

The write circuit of the MTJs in the flip-flop is illustrated in Fig. 7.3(b). In this example, the MTJs are programmed when the wclk_b signal is "0" (low). If the input data is "1" (or high), the output of the upper NOR gate is "1" and that of the lower gate is "0." The write current flows from Qn7 through the write line to Qn4. Since the two MTJs are written with currents (or fields) of different directions, they are written into opposite states. Unlike the programming of MRAM, only a single predominantly easy-axis write field is required to set the MTJ state. Thus, there is only one program line per flip-flop.

The transfer curves of the unbalanced MTJ flip-flop are shown in Fig. 7.4, and they exhibit one metastable and two stable states of different distances from the metastable state. To read the MTJ states into the flip-flop, a read pulse is applied. The read pulse activates transistor Qn3, and the inputs of the two inverters are shorted together. This causes the inverters to be pulled into a metastable state. When the read signal is released, Qn3 is deactivated, and the output nodes Q and Q_b can be moved (to the intersection of curves (2) and (3) in Fig. 7.4) by the two complementary MTJs to one of the two stable states, whichever is closer. For example, if MTJ2 is in a P state and MTJ1 is in a AP state, the output Q moves to "0" (low).

It is worth pointing out that, as long as the read signal is low, the output of the flip-flop remains latched and stays in the same state. One may program the MTJ

(a)

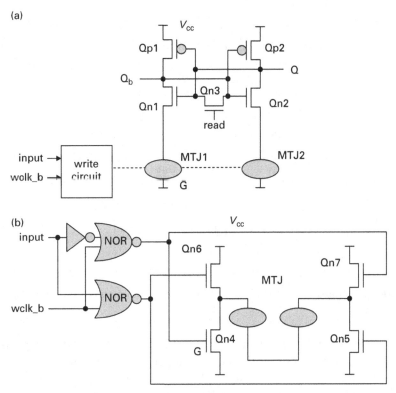

(b)

Figure 7.3. (a) MTJ unbalanced flip-flop. (b) Write circuit of MTJs. (After ref. [8], with permission from IEEE.)

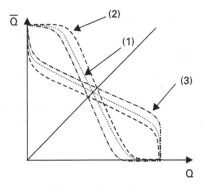

Figure 7.4. Transfer curves of unbalanced MTJ flip-flop. The two metastable points are the intersects of the dashed and dashed-dot transfer curves. (After ref. [8], with permission from IEEE.)

state at any time, even when the flip-flop is running. In other words, the updated MTJs will not change the flip-flop state unless another read signal is applied after the MTJ is updated. Thus, information can be written in the MTJ during runtime. This property makes the unbalanced MTJ flip-flop runtime reconfigurable.

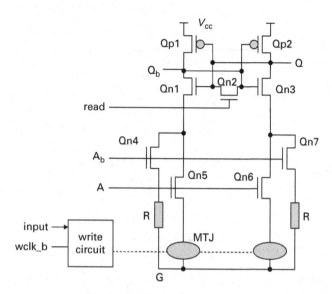

Figure 7.5. Non-volatile multiplexer. (After ref. [8], with permission from IEEE.)

When powering up the unbalanced MTJ flip-flop, the voltage supply is raised to V_{cc}, and the output of the flip-flop is automatically set to a state by the data stored in the MTJ. To reset the output, the MTJs are first re-written into the opposite state and then the flip-flop is updated with a read signal.

7.3.3 Non-volatile multiplexer

Here, we will discuss the functions of the derivatives of the MTJ flip-flop. A multiplexer is shown in Fig. 7.5. Having the source of NMOS connected to either a constant resistor R (by activating signal A_b) or a MTJ (by activating signal A) through the switches, the output of the multiplexer can be either the input data or a fixed value that is pre-determined by the MTJ states. The multiplexer is initialized by the MTJ when being powered up. During its subsequent operation, the MTJ is decoupled from the flip-flop, and the multiplexer acts like a regular flip-flop [8].

7.3.4 MTJ data register

A SRAM-based data register is made up of a master flip-flop and a slave flip-flop circuit. The input data are clocked into the master and the slave latch at the rising and falling clock edges, respectively. To make the flip-flop non-volatile, one of the two flip-flops is replaced with a MTJ flip-flop. Figure 7.6 shows a SRAM-based data register.

Figure 7.7 illustrates one non-volatile data register circuit configuration. The data register is made up of a conventional master flip-flop and a slave MTJ flip-flop. The slave latch is non-volatile and a write clock (wck) activates the

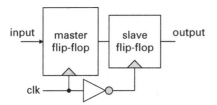

Figure 7.6. Data register.

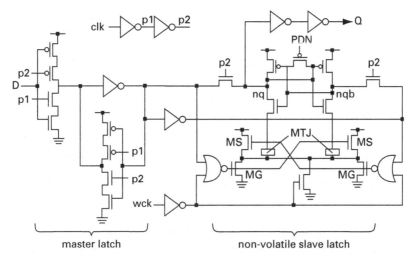

master latch   non-volatile slave latch

Figure 7.7. Non-volatile high-speed data register having a pair of MTJs. (After ref. [7], with permission from IEEE.)

MTJ write circuit. The MTJ is written with the latest data state held by the register. When the write clock is low, the write circuit is deactivated. The slave latch keeps the data when power is interrupted or down. The MTJ pair can be written at an extremely short clock cycle, 3.5 GHz, and at a power supply voltage of 1.2 V.

FeRAM-based non-volatile data registers are already on the market, but, due to their limited write endurance and the requirement of high write voltage, their application is limited.

The magnetic flip-flop currently finds applications in the reduction of SoC chip power and runtime reconfiguration of systems, such as the FPGA (field-programmable gate array) [8], and CAM (content-addressable memory) (for a tutorial, see ref. [9]). There may be other system function level applications that are yet to be discovered.

7.4   System-on-chip power reduction

System-on-chip (SoC) circuit power consumption tends to increase in relation to miniaturization and greater circuit sophistication. However, the demand for

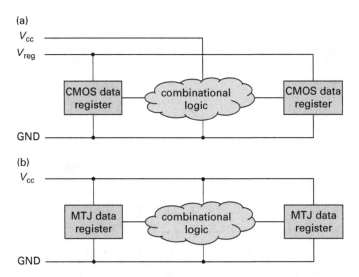

Figure 7.8. A logic circuit can be partitioned into registers and combinational logic. At the end of each clock cycle, data from the combinational logic are saved in the register. (a) The power supply to the combinational logic is V_{cc} and the power supply to the registers is V_{reg}. (b) With non-volatile registers, a single supply is enough, and the entire SoC can be powered down.

electronic appliances that use SoC circuits and consume low power is steadily increasing. It has now become an essential task for electronic appliance development to address the need for both advanced functions and low power consumption. The development of mobile appliances powered by batteries is particularly influenced in consideration of their need to reduce power consumption in both standby and operational modes.

In deep sub-micron CMOS circuits, the dc power dissipation is frequently as large as the ac power, due to the large leakage current from the drain to the source of the transistor. The most popular approach taken to lower the chip power is to lower the power supply voltage V_{cc} in the portion of the logic circuit that is in standby. One problem of this low V_{cc} approach is that the signal-to-noise ratio of the registers drops and the registers become less stable and do not reliably hold data. The present solution to such a problem is to have two sets of voltage supply. One supplies the combinational logic circuit and the other supplies the registers. Figure 7.8(a) illustrates the power supply scheme. During standby, the voltage of the combinational circuit V_{cc} is lowered to save power, while the voltage of the registers V_{reg} is kept unchanged in order to save the data in the registers. Thus, the power of the registers, which is a substantial fraction of the chip power, is hardly reduced.

To shut down the power of the entire chip completely, one needs to copy the data in all the registers to a non-volatile memory before shut down. The same data are reloaded back to the registers after power-up before the chip can resume operation. Each operation of this type can take a long time. For example, to store data from 10^4 registers into a RAM sequentially, each write access of the RAM takes $100\,\mu s$ (as in FLASH memory); the total time will be 1 s. That will waste 10^9 cycles for a 1 GHz clock CPU! When powering up, the same amount of time is required to reload the data back into the registers, and the system must wait for the completion of data loading prior to resuming operation. Thus, the performance penalty is very significant. In addition, a non-volatile memory, which may be on the same chip or off chip, is needed to save the data in the registers.

Having MTJ data registers in the logic chip, as illustrated in Fig. 7.8(b), will allow one to shut down the power of the entire chip immediately and resume operation right after power-up with little performance penalty.

7.5   Runtime reconfigurable electronic system

The main elements used in the FPGA are configurable logic block (CLB), look-up tables (LUT) and interconnect between these blocks. In many cases, SRAMs are used in the switching matrix to drive the pass gate transistor. The MTJ flip-flop and its derivatives can be used to replace the LUT unit, and the MTJ multiplexer can be used to replace the SRAM cells in the cross-point switches. As a result, the circuit complexity (or the number of transistors) is reduced, and the power-up initialization procedure is eliminated since the MTJ flip-flop is non-volatile [8].

References

[1] D. R. Baselt, G. U. Lee, M. Natesan, S. W. Metzger, P. E. Sheehan and R. J. Colton, *Biosensors & Bioelectron.* **13**, 731 (1998).

[2] C. R. Tamanaha, S. P. Mulvaney, J. C. Rife and L. J. Whitman, *Biosensors & Bioelectron.*, doi:10.1016/j.bios.2008.02.009 (2008).

[3] S. J. Osterfelda, H. Yub, R. S. Gaster *et al.*, *Proc. Natl Acad. Sci.* **105**(52), 20637 (2008).

[4] N. Sakimura, H. Honjo, S. Saito *et al.*, *IEEE Digest of Technical Papers, ISSCC* (2009), paper 27.4.

[5] Y. Iwata, K. Tsuchida, T. Inaba *et al.*, *IEEE Digest of Technical Papers, ISSCC* (2006), p. 138.

[6] N. Sakimura, T. Sugibayashi, T. Honda *et al.*, *IEEE J. Solid-State Circ.* **42**(4), (2007).

[7] N. Sakimura, T. Sugibayashi, R. Nebashi and N. Kasai, *IEEE 2008 Custom Integrated Circuit Conference (CICC)* September 21–24 (2008), paper 14–4, p. 325.

[8] N. Bruchon, L. Torres, G. Sassatelli and G. Cambon, *IEEE Proceedings of the 2006 Emerging VLSI Technologies and Architectures (ISVLSI'06)*, Karlsruhe, Germany, March 2–3 (2006).

[9] K. Pagiamtzis and A. Sheikholeslami, *IEEE J. Solid-State Circ.* **41**(3), 712 (2006).

Appendix A   Unit conversion table for cgs and SI units

Force, F	1 dyne (dyn)	10^{-5} newton (N)
Magnetic field strength, H	1 oersted (Oe)	79.58 ampere/meter (A/m)
Magnetic induction, B	1 gauss (G)	10^{-4} tesla (T)
Electric field strength, E	1 erg	10^{-7} joule (J)
Magnetic flux density, Φ	1 maxwell	10^{-8} weber (Wb)
Magnetization, M	1 emu/cm^3	12.57×10^{-4} Wb/m$^2 = 12.57 \times 10^{-4}$ A/m

Appendix B Dimensions of units of magnetism

SI magnetic units are easily related to the current, voltage and energy in MKS units, since the SI system was originally developed under the assumption that magnetism is originated from electric current. The dimensions of magnetic units are shown below; A = amperes, s = seconds, kg = kilograms and m = meters.

(1) newton (N) = $kg\, m/s^2$;
(2) joule (J) = $kg\, m^2/s^2$;
(3) magnetic field (H) = A/m;
(4) henry (h) = $kg\, m^2/s^2 A^2$;
(5) tesla (T) = $kg/s^2 A$;
(6) weber (Wb) = $kg\, m^2/s^2 A$.

Magnetism in cgs units is less transparent. The unit of magnetic moment m is the emu. The density of the magnetic moment M_S is emu/cm^3. The magnetic induction is given by $B = H + 4\pi M_S$, where the magnetic field H is given in units of oersted (Oe) and B is given in gauss (G). (1 Oe = 1 G in air, since $M = 0$ in air.) Thus, the "4π" in $4\pi M_S$ is not dimensionless. Its dimension is [G/(emu/cm^3)], and it is equivalent to the inverse of susceptibility, so that the unit of $4\pi M_S$ is [G/(emu/cm^3)] · [emu/cm^3] = G.

The dimension of "emu" can be understood from the dimension of energy. In Chapter 2, we discussed the magnetostatic energy per unit volume of magnetic material under a magnetic field H to be $\sim M_S \cdot H$. The dimension of $M_S \cdot H$ is [emu · cm^{-3}] · [Oe], which should be equivalent to [erg · cm^{-3}]. Therefore, the dimension of emu is [erg/Oe] or [emu/G] in air. From this, we find that the dimension of the volume susceptibility of the "4π" is [G^{2}cm^3/erg].

Recently, many publishers have begun to give A/m as the unit of magnetization, M_S. For a material with $M_S = 1300$ emu/cm^3 (say, cobalt), it is equivalent to 1300 kA/m.

Appendix C   Physical constants

Symbol	Value (unit)
μ_B (Bohr magneton)	9.27×10^{-21} erg/Oe, 9.27×10^{-24} J/T
q (elementary charge)	1.602×10^{-19} C
k_B (Boltzmann constant)	1.38×10^{-34} J/K
k_BT/q (thermal voltage at 300 K)	0.0259 V
h (Planck constant)	6.626×10^{-24} J s
$\hbar$ $(= h/2\pi$, reduced Planck constant)	1.054×10^{-24} J s
ε_0 (permittivity in a vacuum)	8.86×10^{-14} F/m
μ_0 (permeability in a vacuum)	1.257×10^{-6} H/m

Appendix D   Gaussian distribution and quantile plots

Statistically, the random fluctuation of device parameters can be described as a Gaussian (Normal) distribution in the following form:

$$y(x) = \frac{1}{\sqrt{2\pi}\sigma} \exp\left(-\left(\frac{x - x_0}{\sqrt{2}\sigma}\right)^2\right), \tag{D.1}$$

where x_0 is the center of the distribution (called the mean value) and σ is the standard deviation of the distribution. The maximum value of the Gaussian distribution is at $x = x_0$, where

$$y(x_0) = \frac{1}{\sqrt{2\pi}\sigma}.$$

The cumulative distribution of the Gaussian distribution is an error function,

$$erf(x) = \frac{2}{\sqrt{\pi}} \int_{x_0 - x}^{x + x_0} y(t)\, dt. \tag{D.2}$$

When $x = \infty$, the integration covers the entire distribution and $erf(\infty) = 1$. For finite x values, the "escape" of a Gaussian distribution is the *complement error function*, or $erfc(x) = 1 - erf(x)$. It is convenient to describe x in units of sigma, $n\sigma$, and the n value for various size of memory array is shown in Fig. D.1.

Since the Gaussian distribution is a bell-shaped curve on a linear scale plot, its distribution can be re-plotted in a different form: a "quantile plot" in a straight line,

$$q(x) = \left[\ln\left(\sqrt{2\pi}\sigma y(x)\right)\right]^{0.5} = -\frac{(x - x_0)}{\sqrt{2}\sigma}. \tag{D.3}$$

Equation (D.3) shows the Gaussian distribution and its corresponding quantile plot for $x_0 = 0$. The scale on the quantile plot is n, the multiplier of sigma. For 6σ, one covers the distribution to nearly 10^{-8} parts of the Gaussian tail.

In an ensemble of data points $\{y_n\}$ that fall into the bimodal Gaussian distribution, the quantile plot is shown in Fig. D.2(b). The quantile plot provides a simple

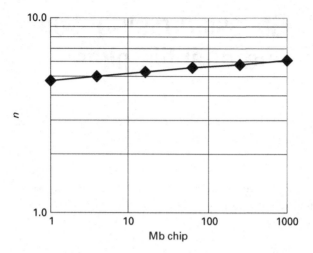

Figure D.1. The number of sigmas needed to cover all the bits in the entire array of a Mb-level chip.

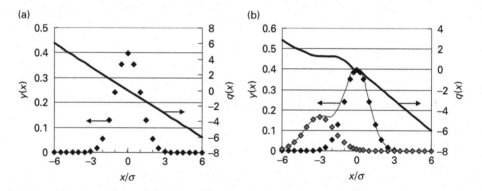

Figure D.2. (a) Gaussian (or Normal) distribution and the quantile plot. (b) Bimodal Gaussian distribution and the corresponding quantile plot ($x_0 = 0$, -3; amplitude ratio $= 1 : 0.4$; $\sigma = 1$, 1.2).

means to separate single and bimodal Gaussian distributions and to project the spread using limited input data. For example, one can project ppm (parts per million) level defects by examining defects in only 1000 parts.

Appendix E   Weibull distribution

The probability density function of the general Weibull distribution is

$$f(x) = \frac{\beta}{\alpha}\left(\frac{x-\mu}{\alpha}\right)^{(\beta-1)} \exp\left(-\left(\frac{x-\mu}{\alpha}\right)^{\beta}\right), \; x \geq \mu; \; \beta, \alpha > 0, \tag{E.1}$$

where β is the shape parameter, μ is the location parameter and α is the scale parameter. For $\mu = 0$ and $\alpha = 1$, $f(x)$ is called the standard Weibull distribution. For $\mu = 0$, $f(x)$ is called the two-parameter Weibull distribution. The equation for the standard Weibull distribution reduces to

$$f(x) = \frac{\beta}{\alpha}(x/\alpha)^{(\beta-1)} \exp\left(-(x/\alpha)^{\beta}\right), x \geq 0; \; \beta, \alpha > 0. \tag{E.2}$$

The accumulative distribution function of the Weibull distribution is given by

$$F(x) = 1 - \exp\left(-(x/\alpha)^{\beta}\right), \; x \geq 0; \; \beta, \alpha > 0. \tag{E.3}$$

The accumulative distribution $F(x)$ starts at 0 and ends at 1. Figure E.1 shows the Weibull distribution and accumulative Weibull distribution functions.

One may manipulate Eq. (E.3) so that, $1 - F(x) = \exp(-(x/\alpha)^{\beta})$. By taking the natural logarithmic function of both sides, one obtains

$$g(x) = \ln\{-\ln[1 - F(x)]\} = \beta \cdot \ln(x/\alpha). \tag{E.4}$$

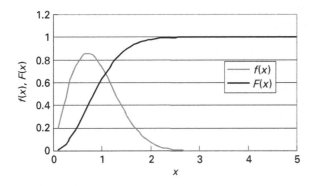

Figure E.1. Weibull distribution $f(x)$ and its cumulative function $F(x)$ for $\beta = 2$ and $\alpha = 1$.

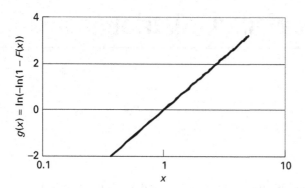

Figure E.2. Weibull plot of the cumulative function of Fig. E.1.

Plotting $g(x)$ vs. ln (x) for several α's, one can obtain the slope β and the α from the intersect at $g(x) = 0$. Figure E.2 shows the Weibull plot of $g(x)$ function with $\alpha = 1$. It is a straight line with slope 2 and intersects with $g(x) = 0$ at 1; thus $\alpha = 1$.

Appendix F Time-dependent dielectric breakdown (TDDB) of magnetic tunnel junction devices

The dielectric breakdown characteristics of magnetic tunnel junctions are the primary reliability concern. It has been found that the distribution of time-dependent dielectric breakdown (TDDB), t_{BD}, as a function of stress voltage across a MTJ, V_{MTJ}, closely follows the Weibull distribution [1]:

$$F(x) = 1 - \exp(-(x/\alpha)^\beta).$$

In this case, x is the time-to-breakdown, t_{BD}, at a given stress voltage; α is the scaling factor between t_{BD} and the stress voltage, V_{MTJ}, and β is the activation energy – both are extracted from the Weibull plot. The operating life of a MTJ at a given V_{MTJ} can therefore be projected by constructing a series of Weibull plots of the TDDB of different V_{MTJ} stress voltages, usually greater than the operating voltage. Thus, the tunnel barrier of the MTJ breaks down in a short time.

By measuring the t_{BD} of a large number of MTJs at a stress voltage V_{MTJ}, one obtains the cumulative distribution of $F(t_{BD})$. From $F(t_{BD})$, one may

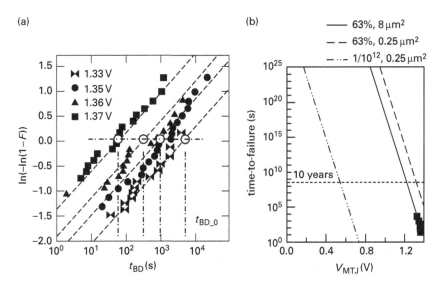

Figure F.1. (a) Weibull plot of time-dependent dielectric breakdown (TDDB) of four different stress voltages; t_{BD_0} is the time at which 63% of the MTJs break down. (b) Time-to-failure vs. V_{MTJ}. (After ref. [1].)

construct one line on the Weibull plot $g(t_{BD}) = \ln(-\ln(1 - F(t_{BD})))$. Repeating the same measurement at a different stress voltage, one can construct another line. Figure F.1(a) shows the Weibull plots of $g(t_{BD})$ of four stress voltages. The slope of the $g(t_{BD})$ data line of each stress voltage is roughly the same. For each stress voltage V_{MTJ}, 63% of the devices break down at a time that each curve intersects at $g(t_{BD\_0}) = 0$. Knowing β, one can project the operating voltage of the MTJ at a breakdown probability of 10^{-12}.

Reference

[1] J. Das, R. Degraeve, P. Roussel, G. Groeseneken, G. Borghs and J. De Boeck, *J. Appl. Phys.* **91**(10), 7712 (2002).

Appendix G   Binomial distribution and Poisson distribution

If we know that the average probability of a MTJ switching at a given switching current is 0.1, and we measure ten MTJs, what is the probability that *exactly* three out of ten MTJs switch? This is a classical binomial distribution problem. The probability is given by

$$P(k,n) = \frac{n!}{(n-k)!k!} \lambda^k (1-\lambda)^{n-k}, k = 0,1,2,3,$$

where λ is the average probability of success, and we obtain k, the number of successes, out of a very large number of attempts. The probability of each MTJ switching is $\lambda = 0.1$, that of not switching is $(1-\lambda) = 0.9$. The probability of $k = 3$ out of $n = 10$ MTJs switching (and seven not switching) is

$$P(3,10) = \frac{10!}{7! \cdot 3!} (0.1)^3 \cdot (0.9)^{10-3}$$

i.e.

$$P = 5.7\%.$$

The probability of one out of ten MTJ switching is much higher, 40%. The probability of the number of MTJs that switch is shown in Fig. G.1.

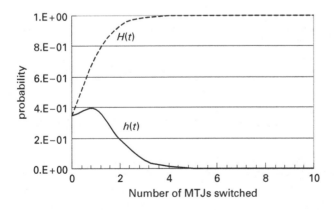

Figure G.1. Binomial distribution (solid line) of the number of successes out of ten attempts and its cumulative function (dashed line). The average success rate is 0.1.

When the attempt number n is very large, say approaching infinity, the binomial distribution approaches a Poisson distribution given by

$$h(\lambda, k) = \frac{\lambda^{-k} \exp(-\lambda)}{k!}.$$

The cumulative function of the Poisson distribution is

$$H(\lambda, x) = \sum_{k=0}^{x} \frac{\lambda^{-k} \exp(-\lambda)}{k!}.$$

Appendix H Defect density and the breakdown/TMR distribution of MTJ devices

The probability of n events, given a mean rate of λ is

$$P(n, \lambda) = \frac{\lambda^n}{n!} \exp(-\lambda). \tag{H.1}$$

The average number of defects (the "rate," as above) $\lambda = d \cdot A$, where d is the defect density (number of defects/μm^2) and A is the MTJ area. The probability that a bit cell with area A has zero defects is

$$P(n = 0, \ d \cdot A) = \exp(-d \cdot A). \tag{H.2}$$

We postulate that bit cells with no defects exhibit hard (intrinsic) breakdown and bit cells with defects exhibit soft (extrinsic) breakdown. The population of soft breakdown increases as the defect number increases.

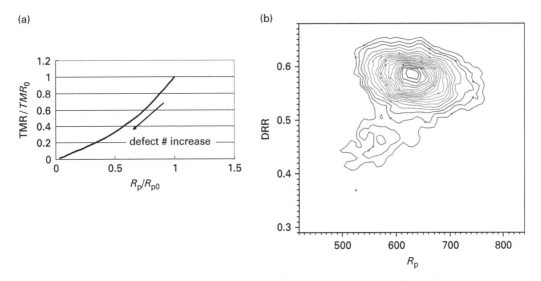

Figure H.1. (a) TMR vs. R_p affected by defects from this model. (b) Measured distribution of TMR vs. R_p. (Courtesy of Robert Beach, MagIC Technologies, Inc.)

Similarly, a defect in a MTJ is considered as a finite resistance, which does not contribute to the TMR action. So, a MTJ with n defects may be modeled as a MTJ in shunt with n fixed resistances:

$$1/R_L = 1/R_p + n/R_{defect}. \tag{H.3}$$

Thus, if $x = n \cdot R_p/R_{defect} \ll 1$, then $R_L \approx R_p(1-x)$. Similarly, based on the same argument, it follows that

$$\text{TMR} = \frac{R_H}{R_L} - 1 \approx TMR_0[1 - (x+x') \cdot (TMR_0 + 1)], \tag{H.4}$$

where $x' = n \cdot R_{ap}/R_{defect}$. Both R_p and the TMR drop off when the number of defects increases. Figure H.1 illustrates the effects of measured TMR vs. R_p.

Appendix I   Fe, Ni and Co material parameters

	Nickel (Ni)	Cobalt (Co)	Iron (Fe)
Atomic number	28	27	26
Lattice structure	face-centered cubic	hexagonal	body-centered cubic
Lattice constant (nm)	0.352	$a = 0.251$, $c = 0.407$	0.287
Group, period, block	10, 4, d	9, 4, d	8, 4, d
Standard atomic weight (g/mole)	58.69	58.93	55.85
Electrons per shell 3(s,p,d), 4s	2, 6, 8, 2	2, 6, 7, 2	2, 6, 6, 2
Saturation magnetization (G)	485	1400	1707
Curie temperature (K)	631	1400	1043
Bohr magneton/atom (0 K)	0.606	1.72	2.22
Density, g/cm^3	8.91	8.9	7.87
Melting point temperature	1728	1768	1811
Electrical resistivity (nΩ m)	69.3	62.4	96.1
Thermal conductivity (W/m K)	90.9	100	80.4
Thermal expansion coefficient (μm/m K)	13.4	13	11.8
Sound speed (m/s)	4900	4720	5120
Young's modulus (GPa)	200	209	211
Shear modulus	76	75	82
Bulk modulus	180	180	170
Poisson ratio	0.31	0.31	0.29

Appendix J   Soft error, hard fail
and design margin

Soft fail is different from hard fail, which results from defects. A soft fail happens intermittently when a cell is written and/or read, and is not repeatable. By testing the memory array repeatedly, and collecting the frequency of error events, one may separate the hard defects from the soft fails. Figure J.1 illustrates the soft and hard fails. Each cell of a 1 Mb cell MRAM array was written and read back 100 times. The test is carried out for four different write currents, i_W. Some cells fail once out of 100 tests; some fail more often.

The write current affects the fail rate. At a write current of 13 mA, more than 100 cells fails once, 30 cells fail twice, and so on. At lower currents, the fail cell count is lower. Nonetheless, fewer cells fail with higher frequency. This kind of cell fail is considered as soft fail.

To the right-hand side of the chart, there are more than 13 cells that failed all 100 tests, at all write currents. These cells exhibit hard fail, most likely caused by defects.

The soft fail is dealt with by introducing a statistical design margin. A single-device parameter margin model is shown in Fig. J.2. The relevant device parameter is assumed to follow a Gaussian distribution

$$f(x) = \frac{N}{\sqrt{2\pi}\sigma} \exp\left(-\left(\frac{x - x_0}{\sqrt{2}\sigma}\right)^2\right),$$

where N is the total number of samples, x_0 is the mean value of x and σ is the sigma. The fail rate is determined by the distance between x and a fail reference x_F in the form of

$$P_F(x) = P_{F0} \exp\left(-\left(\frac{x - x_F}{\sqrt{2}\sigma_F}\right)^2\right) \quad \text{for} \quad x < x_F,$$

and

$$P_F(x) = 1 \quad \text{for} \quad x \geq x_F.$$

Thus, the statistical fail count is given by

$$Fail\_Count = \int_{-\infty}^{\infty} f(x) P_F(x)\, dx. \tag{J.1}$$

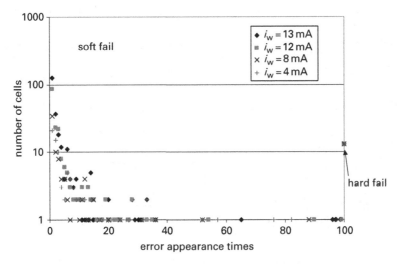

Figure J.1. An example of soft and hard fail bits in a MRAM cell array. Cells are repeatedly written then read 100 times. The number of fail cells is plotted vs. the frequency of failure appearance. (Courtesy of MagIC Technologies, Inc.)

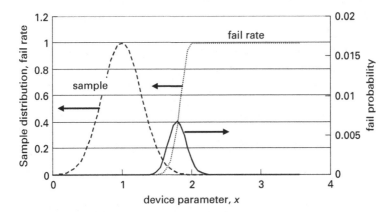

Figure J.2. Sample parameter distribution, fail rate and sample fail probability of single-device parameter model. Both sample distribution and *Fail_Count* are normalized by sample size N. In this example, $x_0 = 1$, $x_F = 2$.

In the memory, the bit cell failure may result from write fail (write unsuccessful) or read fail (read incorrectly). In a MRAM, soft write fail occurs when the write current is marginal; the bit cell can sometimes be written successfully, but not always. The write disturb is another kind of soft fail.

The read soft fail may happen to a bit cell when the amplitude of the read back signal barely overcomes the sense amplifier noise/mismatch. The same bit cell may be read correctly in certain time and incorrectly in others. This kind of intermittent soft fail can only be eliminated by widening the statistical margins.

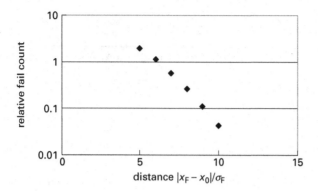

Figure J.Q1. Number of cells vs. fail times in 100 repeated tests. Those that fail 100 times are considered as hard fail; the rest are soft fail.

Homework

QJ.1 Let the device parameter distribution be Gaussian, and let $x_0 = 1$ and $\sigma = 0.2$. The σ_F of the fail distribution is 0.1. Calculate the fail count when the distance $|x_0 - x_F|$ is from 1 to 0.5.

AJ.1 The *Fail_Count* is the integration of the fail probability, which is the area under the fail propability curve of Fig. J.2. The fail count decreases nearly exponentially with the distance $|x_0 - x_F|/\sigma_F$ as shown in Fig. J.Q1.

Index